Lecture Notes in Control and Information Sciences

Edited by M. Thoma and A. Wyner

For information about Vols. 1-96 please contact your bookseller or Springer-Verlag

Lecture Notes in Control and Information Sciences

Edited by M. Thoma and A. Wyner

164

I. Lasiecka, R. Triggiani

Differential and Algebraic Riccati Equations with Application to Boundary/Point Control Problems: Continuous Theory and Approximation Theory

Springer-Verlag Berlin Heidelberg GmbH

Authors
Prof. Irena Lasiecka
Prof. Roberto Triggiani

Dept. of Applied Mathematics
Thornton Hall
University of Virginia
Charlottesville, VA 22903
USA

ISBN 978-3-540-54339-8   ISBN 978-3-540-47564-4 (eBook)
DOI 10.1007/978-3-540-47564-4

Preface

These Lecture Notes collect, in a unified framework, an updated
and rather comprehensive account of results centered on the theory of
optimal control with quadratic cost functionals for abstract (linear)
equations in a Hilbert space Y, of the type $\dot{y} = Ay+Bu$. Here, A (free
dynamics operator) is at least the generator of a s.c. (strongly
continuous) semigroup and B (control operator is an unbounded operator
with a degree of unboundedness up to the degree of unboundedness of A.
Also, u is the control function which is L_2 in time with values in a
Hilbert space U. The present treatment includes both the finite as
well as the infinite time horizon problems. It culminates with the
analysis of the corresponding differential and algebraic Riccati
(operator) equations, which arise in the (pointwise) feedback synthesis
of the optimal solution pair $\{u^0, y^0\}$. These Notes give the main
results of the theory--which has reached a considerable degree of
maturity over the past few years--as well as the authors' basic
philosophy of approach, which is contained in their work of the past
fifteen years. Only some key points of the technical developments are
provided, while, for the most part, detailed technical proofs are
referred to in the appropriate literature. Both continuous theory as
well as numerical approximation theory for the Riccati equations are
included.

Essentially, two (non-necessarily mutually exclusive) classes of
abstract equations are identified by means of a respective abstract
assumption, (H.1) and (H.2) below. These assumptions are, in fact,
nothing but properties which capture distinctive features of the
concrete classes of partial differential equations of interest.

(1) First class. This includes parabolic-like dynamics: not only parabolic (or diffusion) equations but also wave and plate equations with a high degree of internal damping. All these equations are identified by the property that the free dynamics operator A generates a s.c. analytic semigroup on Y.

(2) Second class. This includes wave-like (hyperbolic) dynamics, plate-like (both hyperbolic or not) dynamics, and Schrödinger equations. These are identified by a distinctive abstract 'trace regularity' property of the corresponding free dynamics (and, by duality, an 'interior regularity' property of the corresponding non-homogeneous problem).

In either case, the control operator B may have, as remarked already, a degree of unboundedness up to the degree of unboundedness of the free dynamics operator A. This framework captures, among others, mixed problems for partial differential equations.

Special emphasis is paid to the following topics.

(i)   Abstract operator models for boundary control problems.

(ii)   Identification of the space Y of optimal regularity of the solutions, particularly for the second class: it is with respect to the norm of this space that then the solution y is penalized in the quadratic cost functional.

(iii)   Identification of the regularity properties of the optimal pair $\{u^0, y^0\}$.

(iv)   Verification of the so-called 'finite cost condition' (F.C.C.) in the space Y (of optimal regularity mentioned in (ii)), in the case of the infinite time horizon problem and related algebraic Riccati equations. This is the property whereby for each initial condition in Y, there exists some $u \in L_2(0,\infty;U)$ such that the corresponding solution $y \in L_2(0,\infty;Y)$ so that the quadratic cost functional is finite.

(v)   Constructive variational approach to the issue of existence of a
        solution (Riccati operator) of a Riccati equation, whether
        differential or algebraic.

(vi)   Development of numerical algorithms which reproduce numerically
        the key properties of the continuous problems.

        (i) As to the abstract modeling problem (i), the treatment
relies, in the parabolic case, on the ideas of [B.1, Sect. 4.12], [W],
as simplified, and refined in [T.5], [T.6], [Las.4], [L-T.5], by means
of elliptic theory and identification [Gr.1] between Sobolev spaces and
domains of appropriate fractional powers of the basic differential
operator A; and in the hyperbolic/plate case, on the model ideas of
[T.2], [L-T.1]. These works benefited from and pushed further to use
an incipient idea of [Fa.1]. See Notes at the end of Section 4 in
[L-T.24] for more details on operator modeling. These operator models
have been successfully used by the authors in a large variety of
boundary control problems (optimal quadratic cost problems and Riccati
equations; uniform stabilization; spectral properties assignment and
stabililization via a feedback operator, etc.).

        (ii) As to the optimal regularity problem (ii), we point out a
drastic difference in the role played by abstract models in
hyperbolic/plate mixed problems (second order in time) or Schrödinger
equations on the one hand, and parabolic mixed problems (first order in
time) on the other. In the latter case, the combination of semigroup
methods with elliptic theory and identification of domains of
appropriate fractional powers with Sobolev spaces is sufficient to
provide (or re-prove) optimal regularity results for parabolic mixed
problems, see e.g., [Las.4] and [L-T.4], [L-T.8]. This theory includes
the 'Hilbert theory' of, say, [Lio-Mag.1], which is obtained by
different (energy) means. Not so, however, for hyperbolic/mixed

problems. Here, the first crucial step or building block of a
regularity theory comes from purely partial differential equation
methods (energy, or multiplier methods, either in differential form or
else in pseudo-differential form) which were brought to bear on these
problems only very recently, beginning with second-order hyperbolic
equations with Dirichlet control [L-T.1], [L-T.2], [Lio.2], [L-L-T].
In the case of hyperbolic/plate/Schrödinger mixed problems, abstract
operator methods provide useful tools only at a subsequent level (for
higher/lower data, duality or transposition, etc.), after a key
preliminary regularity result--typically a trace regularity property of
the corresponding homogeneous problem--has been obtained from energy
methods. These 'trace regularity' properties can then be abstracted
and unified in the 'abstract trace property' (H.2) of the second class.

(iii) As to the regularity properties of the optimal pair
$\{u^0, y^0\}$, these rely on regularity theory with input u smoother than L_2
and, in the parabolic case, on a 'boost-strap' argument.

(iv) As to the Finite Cost Condition (F.C.C.) of problem (iv),
we note that in the case of the first class (parabolic-like dynamics),
the F.C.C. is most readily verified via uniform feedback stabilization,
as the unstable space of the dynamics is at most finite dimensional.
Instead, in the case of the second class (wave/plate/Schrödinger
dynamics), the F.C.C. is verified via a study of the exact
controllability property on the space of optimal regularity as in (ii)
[or, alternatively, via the generally more demanding property of
uniform feedback stabilization through and explicit, dissipative
'velocity' feedback]. The regularity and exact controllability issues
amount to an upper bound and, respectively, a lower bound of a suitable
trace of the homogeneous solution with respect to the initial
conditions. Their verification is obtained by partial differential
equation methods (energy/multiplier methods in differential or

pseudo-differential form [L-T.25] and, in the case of second-order hyperbolic equations, on micro-local analysis [B-L-R] to achieve sharp results.

Another approach is proposed in [Lit.]. A vast literature has emerged on these issues over the past five years.

(v) As to the constructive approach to the existence of a Riccati operator mentioned in (v), this consists in two steps: first, one uses the optimality conditions to construct an explicit candidate Riccati operator, defined in terms of the original data of the problem and subsequently one shows that this candidate is a 'bona fide' Riccati operator, which solves, in fact, the corresponding Riccati equation. This approach was introduced (abstractly) in connection with a parabolic problem with Dirichlet control in [L-T.4] (T < ∞) and since then it has been systematically used by the authors in hyperbolic problems [L-T.5], [L-T.6], [L-T.10] (both T < ∞ and T = ∞); in an abstract treatment for the second class [F-L-T] for T = ∞; and in abstract parabolic-like problems [L-T.22]. In the case of T = ∞, under a suitable 'detectability condition', the optimal quadratic cost theory provides another feedback operator, based this time on the Riccati algebraic operator, which yields uniform stability, an additional bonus of the optimization theory. Uniform stabilization via a Riccati operator generally acts on both position and velocity for second-order problems in time, and need not be dissipative.

As an integral part, these Notes contain also a large collection of illustrative examples of boundary/point control problems for partial differential equations, where all the required assumptions are indeed verified to hold true. This also applies to the numerical analysis of the corresponding numerical schemes. Thus the abstract theory is justified.

These Notes are a substantial outgrowth (approximately in the ratio 3 to 1) of the authors' review article entitled:

Algebraic Riccati equations arising in boundary/point control: A review of theoretical and numerical results. Part I: Continuous theory; Part II: Approximation theory, in Perspectives in Control Theory, Proceedings of the Sielpia Conference, Sielpia, Poland, 1988, Editors: B. Jakubczyk; K. Malanowski; and W. Respondek, Birkhauser, Boston, 1990, pp. 175-235.

Also, the authors' reference [L-T.24] provides a review of relevant regularity theory of second-order hyperbolic equations invoked in the text, and based on [L-T.1], [L-T.2], [Lio.1], [L-L-T] in the Dirichlet case and on [L-T.20], [L-T.21], [L-T.23] in the Neumann case. These notes are presently being expanded into a self-contained book.

Finally, the authors gratefully acknowledge financial support received by the following agencies and institutions for research work reported in these Notes: National Science Foundation, Division of Mathematical Sciences; Air Force Office of Scientific Research, Mathematical and Information Sciences; Consiglio Nazionale delle Ricerche, Italy; Scuola Normale Superiore, Pisa, Italy.

Table of Contents

1. Introduction: two abstract classes; statement of main problems

Problem and preliminary model. Consider the following optimal control problem: Given the dynamical system

$$y_t = Ay+Bu; \quad y(0) = y_0 \in Y \tag{1.1}$$

minimize the quadratic functional

$$J(u,y) = \int_0^T [\|Ry(t)\|_Z^2 + \|u(t)\|_U^2]dt + \|Gy(T)\|_W^2 \tag{1.2}$$

over all $u \in L_2(0,T;U)$, with y solution of (1.1) due to u. Throughout the paper, we shall make the following standing assumptions on (1.1), (1.2):

(i)   Y, U, Z, and W are Hilbert spaces;

(ii) $A: Y \supset \mathcal{D}(A) \rightarrow Y$ is the generator of a s.c. semigroup e^{At} on Y, $t > 0$;

(iii) $B: U \rightarrow [\mathcal{D}(A^*)]'$, the dual of $\mathcal{D}(A^*)$ with respect to the Y-topology, A^* being the Y-adjoint of A; more precisely, it is assumed that

$$(A)^{-\gamma}B \in \mathcal{L}(U;Y) \text{ for some constant } 0 \leq \gamma \leq 1; \tag{1.3}$$

(iv)   the operator G is bounded,

$$G \in \mathcal{L}(Y;W). \tag{1.4}$$

Instead, the operator R will be generally bounded: $R \in \mathcal{L}(Y;Z)$, except in Section 4 where it will be allowed to be unbounded. (The analysis which follows Theorem 5.8 will document that the case where R is unbounded and B is bounded is qualitatively easier than the case where R is bounded and B is unbounded.)

In (1.2), $T > 0$ may be finite or infinite. If $T = \infty$, we shall then take $G \equiv 0$ in (1.2).

Abstract classes. Our motivation comes from, and is ultimately directed to, partial differential equations with boundary or point control (in any space dimensions), including those that may arise in 'flexible structures' problems; see the examples of Section 6 and 7 below. Accordingly, we shall distinguish two general classes of not necessarily mutually exclusive dynamics, for which different treatments must be applied in order to capture optimal properties thereof. Each of the two classes will be singled out and modelled by one of the two ('regularity') assumptions below.

First class. The first class satisfies the assumption:

(H.1) $\left\{\begin{array}{l} \text{the s.c. semigroup } e^{At} \text{ is analytic on } Y, \\ t > 0, \text{ and the constant } \gamma \text{ appearing in} \\ (1.3) \text{ is strictly less than 1: } \gamma < 1. \end{array}\right.$ (1.5)

Second class (first form). The second class (first form) satisfies the assumption: for any $0 < T < \infty$, the operator $B^*e^{A^*t}$ admits a continuous extension denoted henceforth by the same symbol satisfying $B^*e^{A^*t}$: continuous $Y \to L_2(0,T;U)$; i.e.,

(H.2)
$$\int_0^T \|B^*e^{A^*t}y\|_U^2 dt \leq c_T\|y\|_Y^2, \qquad y \in Y. \qquad (1.6)$$

In (H.2) we have $B^* \in \mathcal{L}(\mathcal{D}(A^*),U)$ for the dual of B, after identifying $[\mathcal{D}(A^*)]'$ with $\mathcal{D}(A)$

$$(B^*v,u)_U = (v,Bu)_Y; \quad v \in \mathcal{D}(A^*), \ u \in U. \qquad (1.7)$$

Second class (second form). The second class (second form) satisfies the assumption: for any $0 < T < \infty$, the operator $B^*e^{A^*t}R^*$ admits a continuous extension (denoted henceforth by the same symbol) satisfying $B^*e^{A^*t}R^*$: continuous $Z \to L_2(0,T;U)$:

(H.2$_R$)
$$\int_0^T \|B^*e^{A^*t}R^*z\|_U^2 dt \leq c_T\|z\|_Z^2, \qquad z \in Z, \qquad (1.8)$$

where R^* from Z to Y is the adjoint of R.

While conditions (H.1) and (H.2) involve only the operators A and B, instead condition (H.2$_R$) requires also the (possibly unbounded: in Section 4) operator R, as pointed out by the notation.

When $T = \infty$, we shall need for either class an assumption that guarantees the existence of a unique optimal pair $\{u^0,y^0\}$ of the optimal control problem (1.2)

(F.C.C.) $\begin{cases} \underline{\text{Finite Cost Condition}}: \text{ For every } y_0 \in Y, \\ \text{there exists } u \in L_2(0,\infty;U) \text{ such that the} \\ \text{corresponding functional in (1.2)} \\ \text{satisfies } J(u,y(u)) < \infty. \end{cases}$ (1.9)

<u>Remark 1.1</u>. Condition (1.6) implies (1.8) if R is bounded. Conversely, condition (1.8) implies (1.6) if R is an isomorphism. ∎

When $T < \infty$, we introduce the (input-solution) operator L and its L_2-adjoint L^*:

$$(Lu)(t) = \int_0^t e^{A(t-\tau)} Bu(\tau) d\tau, \qquad (1.10)$$

$$(L^*v)(t) = \int_t^T B^* e^{A^*(\tau-t)} v(\tau) d\tau, \qquad (1.11)$$

where $(Lu,v)_{L_2(0,T;Y)} = (u,L^*v)_{L_2(0,T;U)}$. Then

(a)   Condition (H.2) = (1.6) is equivalent to [L-T.2], [L-T.3],

$$L: \text{ continuous } L_2(0,T;U) \to C([0,T];Y), \qquad (1.12)$$

$$L^*: \text{ continuous } L_1(0,T;Y) \to L_2(0,T;U); \qquad (1.13)$$

(b)   Condition (H.2$_R$) = (1.8) is equivalent to

$$RL: \text{ continuous } L_2(0,T;U) \to C([0,T];Z), \qquad (1.14)$$

$$L^*R^*: \text{ continuous } L_1(0,T;Z) \to L_2(0,T;U), \qquad (1.15)$$

where L^*R^* means $(RL)^*$ when one of the factors is unbounded. ∎

To fix our ideas at the outset, the first class covers parabolic-like boundary problems; not only the usual heat equations/diffusion equations, but also wave-like or plate-like problems with high degree of damping ('structural damping'), see Section 6.3 below. Instead, the second class covers undamped, or conservative, or mildly damped wave-like or plate-like partial differential equations (e.g., with viscous damping) with boundary or point control; or Schrödinger equation problems. We shall refer to (H.2) = (1.6) as to an 'abstract' trace theory property, for this is what it amounts to in partial differential problems. The form $(H.2_R) = (1.8)$ of the second class will be applied (in Section 4) to some *purely boundary* optimal control problems for hyperbolic second order equations with Neumann boundary control and Dirichlet boundary observation, the operator R being in this case the (Dirichlet) trace.

Feedback pointwise synthesis and Riccati equations. Qualitatively (and informally), the main problems of interest in this paper are the following, after we assert the existence and uniqueness of an optimal pair $\{u^0(t,0;y_0),y^0(t,0;y_0\}$ of problem (1.1), (1.2).

(P_1) <u>Case</u> T < ∞: Pointwise (in time) feedback representation (synthesis), via a Riccati operator P(t), of the optimal control u^0 in terms of the optimal solution y^0, such as given by

$$u^0(t,0;y_0) = -B^*P(t)y^0(t,0;y_0), \quad \text{a.e. in } 0 \leq t \leq T, \quad (1.16)$$

where the operator P(t) is a solution of an appropriate Differential Riccati Equation (DRE), formally written as

$$\dot{P}(t) + P(t)A + A^*P(t) + R^*R - P(t)BB^*P(t) = 0,$$
$$0 \leq t < T; \quad (1.17)$$
$$P(T) = G^*G,$$

and to be properly interpreted in a technical sense, described below.

(P_2) <u>Case</u> T = ∞: Pointwise (in time) feedback representation, via a Riccati operator, of the optimal control u^0 in terms of the optimal solution y^0, such as given by

$$u^0(t,0;y_0) = -B^*Py^0(t,0;y_0), \quad \text{a.e. in } t > 0, \quad (1.18)$$

where the operator P is a solution of an appropriate Algebraic Riccati Equation (ARE) formally written as

$$PA + A^*P + R^*R - PBB^*P = 0, \quad (1.19)$$

and to be properly interpreted in a technical sense described below.

(P_3) Numerical approximation of the DRE and the ARE for the computation of the Riccati operators $P(t)$ and P.

Difficulties related to B unbounded. The present paper focuses only on the case where the input operator B is unbounded, as in (1.3), and such as it arises in both point control problems and, above all, the more challenging boundary control problems for partial differential equations.· For B bounded, see e.g., [B.1]. Thus, the paper is based only on the natural "regularity" assumptions (H.1), or else (H.2) (not mutually exclusive) for the dynamics, or its variation (H.2$_R$), and--at least at the outset--on a non-smoothing action of the observation operators R and G. This distinction into two (non mutually exclusive) classes is necessary in order to extract best results from each class. Indeed, these two dynamical classes require very different techniques, as they display peculiarly different properties which escape any meaningful and non-artificial "unification." The unboundedness of the operator B contributes to a number of mathematical difficulties in the study of the Riccati feedback synthesis--as expressed by (1.16) for $T < \infty$ and by (1.18) for $T = \infty$ --of the optimal control problem (1.1), (1.2). These difficulties are present at two general levels. (i) One is the abstract level, which is aimed at a general theory of existence and uniqueness of the solution $P(t)$ to the DRE and P to the ARE and the consequent Riccati synthesis, and which must overcome the difficulty inherent to the gain operators $B^*P(t)$ and B^*P. More precisely, in the case $T = \infty$ of ultimate interest, we shall see that for the class of dynamics modelled by assumption (H.2), the gain operator B^*P is inherently unbounded in the most interesting cases of conservative waves and plates problems; or Schrödinger equation. Thus, the classical arguments with B bounded, or variation thereof, are no longer available, and new approaches must be devised. (ii) A second level of difficulty arises in the verification of the abstract assumptions of regularity and, for $T = \infty$, Finite Cost Condition (1.9), particularly for the class modelled by assumption (H.2), in specific, "concrete" p.d.e. problems. Here, p.d.e. techniques are required, which were brought to bear on these problems only very recently. These techniques succeeded in showing the required optimal regularity results (assumption (H.2)) of several waves/plates/Schrödinger boundary control problems, as well as their exact controllability/uniform stabilization properties, which are needed to verify the Finite Cost Condition for these systems. These regularity/exact controllability/uniform stabilization results will be reviewed in Section 7 in the context of each specific dynamics.

We have already pointed out that for the most interesting and typical dynamics modelled by assumption (H.2) (conservative waves and plates, and Schrödinger equations) the gain operator B^*P is unbounded, see Corollaries 5.4 and 5.5. This conclusion rules out, as inapplicable to these systems, other treatments dealing with unbounded B, which however required additional restrictions (on the observation R, on the finite cost conditions, etc.), which are incompatible for these systems; e.g., the results of [P-S], see Remark 5.3 at the end of Section 5.

In addition, in the most distinctive and challenging analytic problems, such as those in Sections 6.1, 6.2, and most of Section 6.3 below, the constant γ in (1.5) is greater than $\frac{1}{2}$: $\frac{1}{2} < \gamma < 1$. Then, all these systems--which are covered by our results--are likewise excluded from the treatment of [P-S]. See Remark 5.1 below.

Overview. The present paper presents an updated and reasonably complete account of results available in the area of the three aforementioned problems for the case where B is unbounded as described in (1.3). The observation operators R on the trajectory and (when $T < \infty$) G on the final state are, at first, taken to be non-smoothing. (In view of the regularity property (continuity) (1.12) for the class of dynamics subject to assumption (H.2) = (1.6), we may just as well take G = 0 in this case; see more on this in Remark 3.1). For instance, our principal results of existence and uniqueness of a Riccati operator are given in this general setting in Section 2.1 for the DRE (class (H.1)) and in Sections 5 for the ARE (classes (H.1) and (H.2)). Subsequently, we shall allow the operator G to be smoothing for the class (H.1) and correspondingly more regular results will be given in this case (Section 2.2). A more regular set-up for dynamics satisfying the regularity assumption (H.2) = (1.6) will likewise be given in Sections 3.2 and 3.3, after a treatment of the non-smoothing case in Section 3.1. In all cases, all results are illustrated by, and applied to, genuine boundary (point) control problems for partial differential equations, typically in any dimension, where we shall verify all the required assumptions.

For the class of systems satisfying the (analyticity) assumption (H.1), the Riccati theory has reached a considerable level of maturity and completeness in both cases, the differential case when $T < \infty$, and the algebraic case when $T = \infty$. This is also so, because all other mathematical problems which interweave with the Riccati theory (regularity, stabilization, etc.) are also well understood for the class (H.1).

In contrast, the situation is more delicate for the classes of hyperbolic equations, plate-like equations, Schrödinger equations, etc. that fall outside the scope of assumption (H.1). Here, a rich Riccati theory is also available in both cases $T < \infty$ and $T = \infty$, to be sure. But it is particularly in the algebraic case when $T = \infty$ and for the class of dynamics satisfying the trace regularity assumption (H.2) = (1.6) that the corresponding Riccati theory may be considered comprehensive and reasonably complete. Equally important, there are at present many 'concrete' p.d.e. problems which serve as illustrations of the abstract theory, where all of the required assumptions are satisfied. This latter step is far from being a trivial one: it involves such delicate mathematical questions as (optimal/sharp) regularity of solutions to mixed problems, as well as exact controllability/uniform stabilization concepts on spaces of optimal regularity. The issues have been resolved only very recently for many (but not all) mixed problems for hyperbolic equations and plate-like equations, not necessarily hyperbolic, as well as for Schrödinger equations. Indeed, one may say that in order to complete the overall picture in the case $T = \infty$ for the class (H.2), a major task which is left for further investigation is not so much to push further the abstract Riccati theory, but to remove the gap existing at present in a few physically important 'concrete' p.d.e. problems between the space of regularity where assumption (H.2) = (1.6) holds true, and the (smoother) space where the Finite Cost Condition (exact controllability/uniform stabilization) has been ascertained so far. This is a purely p.d.e. problem. Among the dynamics where this undesirable gap exists we cite the wave equation with Neumann control, and the Euler-Bernoulli equation with 'high' boundary operators (e.g., second and third). On the other hand, when $T < \infty$, it is the abstract Riccati theory with R non-smoothing that still encounters some subtle issues that make it, overall, less satisfying than the case $T = \infty$, see Section 3. It is precisely in order to capture best results also in the case of *purely boundary* hyperbolic problems, with boundary control and boundary observation, that the variation (H.2$_R$) of the class (H.2) is introduced. That assumption (H.2$_R$) holds true, along with the other abstract assumptions of Section 4, in the context of the purely boundary hyperbolic problem of Neumann type, is a *critical* consequence of the recent sharp regularity theory for the dynamics [L-T.20], [L-T.23].

Part I of this paper deals with the continuous case. It is followed by Part II which deals with an approximation theory thereof.

2. Abstract Differential Riccati Equation for the first class subject to the analyticity assumption (H.1) = (1.5)

We shall first provide a general result under a minimal assumption on G (Section 2.1), and then a more regular result when G is assumed to be a smoothing operator (Section 2.2). In any case, since $0 < T < \infty$ in this section, we may *assume without loss of generality* (modulo an innocuous translation) *that A is boundedly invertible on* Y: $A^{-1} \in \mathcal{L}(Y)$, *and that the fractional powers* $(-A)^{\Theta}$, $0 < \Theta < 1$, *are well defined.*

2.1. The general case

Complementing (1.10), we shall let L_T be the (unbounded) operator

$$L_T u = \int_0^T e^{A(T-t)} Bu(t)dt \tag{2.1}$$

with densely defined domain $\mathcal{D}(L_T) = \{u \in L_2(0,T;U): L_T u \in Y\}$, which describes the map from the input u to the solution y(T) of (1.1) at time t = T, with $y_0 = 0$. Its adjoint L_T^*, $(L_T u, y)_Y = (u, L_T^* y)_{L_2(0,T;Y)}$ is the closed operator

$$\{L_T^* y\}(t) = B^* e^{A^*(T-t)} y, \quad 0 \leq t \leq T, \ y \in Y. \tag{2.2}$$

Theorem 2.1. [L-T.4], [L-T.22] Let the (densely defined) operator GL_T be closed (or closable), as an operator $L_2(0,T;U) \supset \mathcal{D}(GL_T) \to W$. Then, there exists a unique optimal pair $\{u^0(t,0;y_0), y^0(t,0;y_0)\}$ of problem (1.1), (1.2) with $T < \infty$, explicitly given by

$$-u^0(t,0;x) = \{\Lambda_{OT}^{-1}[L_T^*G^*Ge^{AT}x+L^*R^*R(e^{A^\cdot}x)]\}(t), \qquad (2.3)$$

$$y^0(t,0;x) = e^{At}x+(Lu^0)(t), \qquad (2.4)$$

$$\Lambda_{OT} = I+L^*R^*RL+L_T^*G^*GL_T, \qquad (2.5)$$

with L, L*, defined in (1.10), (1.11), and L_T, L_T^* defined by (2.1), (2.2). Moreover, there exists a non-negative, self-adjoint operator $P(t) = P^*(t) \geq 0$, defined explicitly in terms of the data in (x) = (2.20) below) such that

(i) $$P(t) \in \mathcal{L}(Y;C([0,T];Y)); \qquad (2.6)$$

(ii)   in fact, even more, for $0 \leq \Theta < 1$,

$$\|(-A^*)^\Theta P(t)\|_{\mathcal{L}(Y)} \leq \frac{C_{T\gamma\Theta}}{(T-t)^\Theta}, \qquad 0 \leq t < T; \qquad (2.7)$$

(iii)   for any $0 < \varepsilon < T$,

$$(-A^*)^\Theta P(t) \in \mathcal{L}(Y;C([0,T-\varepsilon];Y)), \qquad 0 \leq \Theta < 1; \qquad (2.8)$$

(iv) $$\|B^*P(t)\|_{\mathcal{L}(Y;U)} \leq \frac{C_{T\gamma}}{(T-t)^\gamma}, \qquad 0 \leq t < T; \qquad (2.9)$$

(v)   for any $0 < \varepsilon < T$,

$$B^*P(t) \in \mathcal{L}(Y;C([0,T-\varepsilon];Y)); \qquad (2.10)$$

(vi)   for each $y_0 \in Y$, the optimal control $u^0(t,0;y_0)$ is given in pointwise feedback form by

$$u^0(t,0;y_0) = -B^*P(t)y^0(t,0;y_0), \qquad 0 \leq t < T. \qquad (2.11)$$

(vii)   The following symmetric relation holds:

$$(P(t)x,y)_Y = \int_t^T (Ry^0(\tau,t;x),Ry^0(\tau,t;y))_Z d\tau+(Gy^0(T,t;x),Gy^0(T,t;y))_W$$

$$+ \int_t^T (B^*P(\tau)y^0(\tau,t;x),B^*P(\tau)y^0(\tau,t;y))_U d\tau, \qquad (2.12)$$

from which the optimal cost of the optimal control problem on
$[\tau,T]$ initiating at the time τ at the point $x \in Y$ is

$$J(u^0(\cdot,\tau;x),y^0(\cdot,\tau;x)) = (P(\tau)x,x)_Y \; ; \qquad (2.13)$$

(viii) for $0 < t < T$, $P(t)$ satisfies the following Differential Riccati
Equation for all $x,y \in \mathscr{D}((-A)^\varepsilon)$, $\forall \varepsilon > 0$,

$$(\dot{P}(t)x,y)_Y = -(Rx,Ry)_Z-(P(t)x,Ay)_Y-(P(t)Ax,y)_Y$$

$$+ (B^*P(t)x,B^*P(t)y)_U \; . \qquad (2.14)$$

(ix) The following regularity properties hold true for the optimal
pair

$$\|u^0(\cdot,\tau;x)\|_{L_2(\tau,T;U)}+\|y^0(\cdot,\tau;x)\|_{L_2(\tau,T;U)} \leq c_T\|x\|_Y \qquad (2.15)$$

$$\|Gy^0(T,\tau;x)\|_W \leq c_T\|x\|_Y; \qquad (2.16)$$

$$\|u^0(\cdot,\tau;x)\|_{C_\gamma([\tau,T];U)} \leq c_{T\gamma}\|x\|_Y \qquad (2.17)$$

$$\|y^0(\cdot,\tau;x)\|_{C([\tau,T];Y)} \leq c_{T\gamma}\|x\|_Y \quad \text{if } 0 \leq \gamma < \tfrac{1}{2}; \qquad (2.18a)$$

$$\|y^0(\cdot,\tau;x)\|_{C_{2\gamma-1+\varepsilon}([\tau,T];Y)} \leq c_{T\gamma}\|x\|_Y \quad \text{if } \tfrac{1}{2} \leq \gamma < 1. \qquad (2.18b)$$

In (2.17), (2.18), if X is a Hilbert space and r any real
number, $C_r([\tau,T];X)$ denotes the Banach space defined by

$$C_r([\tau,T];X) = \{f(t) \in C([\tau,T);X): \|f\|_{C_r([\tau,T];X)}$$

$$= \sup_{\tau \leq t < T} (T-t)^r\|f(t)\|_X < \infty\}. \qquad (2.19)$$

Moreover, for $x \in Y$ and for each τ fixed, $0 \leq \tau < T$, the
optimal control $u^0(t,\tau;x)$ and the optimal solution $y^0(t,\tau;x)$
are respectively U-valued and Y-valued functions which are
differentiable in $t \in (\tau,T)$ with $\frac{\partial u^0}{\partial t}(t,\tau;x) \in U$,
$\frac{\partial y^0}{\partial t}(t,\tau;x) \in Y$. In fact, these U-valued and Y-valued
functions $u^0(t,\tau;x)$ and $y^0(t,\tau;x)$ are analytic in $t \in (s,T)$
if the operator A has compact resolvent in Y.

(x)   The operator P(t) is given (constructively) by

$$P(t)x = \int_{t}^{T} e^{A^*(\tau-t)} R^* R y^0(\tau,t;x)d\tau + e^{A^*(T-t)} G^* G y^0(T,t;x). \quad (2.20)$$

(xi)   If we define the evolution operator

$$\Phi(t,\tau)x = y^0(t,\tau;x), \qquad (2.21)$$

the following weak convergence results hold true:

$$\lim_{t\uparrow T}(G\Phi(T,t)x,z)_Z = (Gx,z), \quad \forall\ x \in X,\ \forall\ z \in Z; \quad (2.22)$$

$$\lim_{t\uparrow T}(P(t)x,y)_Y = (G^*Gx,y), \quad \forall\ x,y \in Y. \quad \blacksquare \quad (2.23)$$

Remark 2.1. With reference to the assumption of Theorem 2.1, we have

closed operator $(GL_T)^*$ be		densely defined operator GL_T
densely defined as an operator	$\Longleftrightarrow$	be closable as an operator
$W \supset \mathcal{D}((GL_T)^*) \to L_2(0,T;U)$		$L_2(0,T;U) \supset \mathcal{D}(GL_T) \to W \quad (2.24)$

$L_T^* G^*$ be densely defined

$(-A^*)^{\beta/2}G^*$ be densely defined as
an operator $W \supset \mathcal{D}((-A^*)^{\beta/2}G^*) \to Y$
for some $\beta > 2\gamma-1$

$$(2.25)$$

The equivalence is a standard result [K.1, p. 168]. To see the
sufficient condition, we compute from (2.2),

$$\{L_T^* G^* z\}(t) = B^* e^{A^*(T-t)} G^* z = B^*(-A^*)^{-\gamma}(-A^*)^{\gamma-\beta/2} e^{A^*(T-t)}(-A^*)^{\beta/2}G^* z,$$

$$(2.26)$$

use (1.3), and notice that $(-A^*)^{\gamma-\beta/2} e^{A^*(T-t)} \in \mathcal{L}(Y;L_2(0,T;Y))$ for

$2\gamma-\beta < 1$.

We emphasize that condition (2.25) on $(-A^*)^{\beta/2}G^*$ --which does
not involve B--is only sufficient for the ultimate requirement that GL_T
be closable, which instead involves B. This will be seen in one
example in Section 2.3.

Remark 2.2. (i) An example in Section 2.3 will show that the assumption that GL_T be closed (closable) cannot be dispensed with. ∎

2.2. The smoothing case

In this subsection, in addition to GL_T being closed (closable), we shall assume that G is a smoothing operator in the sense that

$$(-A^*)^\beta G^* G \in \mathcal{L}(Y), \quad \text{for some } \beta > 2\gamma - 1 \tag{2.27}$$

(which is automatically satisfied with $\beta = 0$ if $0 \le \gamma < \frac{1}{2}$). Then, accordingly, stronger results follow. In particular the solution to the DRE (2.14) is unique and the limits as $t \uparrow T$ of Theorem 2.2 are strong.

Theorem 2.2. [D-I], [L-T.22] Assume (2.27). Then:

(i) (Regularity of optimal pair) For $x \in Y$ and any $\varepsilon > 0$,

$$|u^0(\cdot,\tau;x)|_{C_{1-\gamma-\varepsilon}([\tau,T];U)} + |y^0(\cdot,\tau;x)|_{C([\tau,T];Y)} \le c_{T\gamma}|x|_Y, \tag{2.28}$$

$$y^0(T,\cdot;x) = \Phi(T,\cdot)x \in C([\tau,T];Y), \tag{2.29}$$

from which in particular

$$\lim_{t \uparrow T} \Phi(T,t)x = x, \quad x \in Y; \tag{2.30}$$

(ii) for any $0 \le \Theta < 1$, $\varepsilon > 0$, $x \in Y$, we have $(-A^*)^\Theta P(t)x \in C_{\Theta+1-2\gamma+\varepsilon}([0,T];Y)$

$$|(-A^*)^\Theta P(t)|_{\mathcal{L}(Y)} \le \frac{c_{T\gamma}}{1-\Theta} \frac{1}{(T-t)^{\Theta+1-2\gamma+\varepsilon}}; \tag{2.31}$$

(iii) $B^* P(t) \in \mathcal{L}(Y;C_{1-\gamma-\varepsilon}([0,T];U))$, i.e., \tag{2.32}

$$|B^* P(t)x|_U \le \frac{c_T}{1-\gamma} \frac{1}{(T-t)^{1-\gamma-\varepsilon}} |x|_Y; $$

(iv) $$\lim_{t \uparrow T} P(t)x = G^* Gx, \quad x \in Y; \tag{2.33}$$

(v) (uniqueness) the solution P(t), given constructively by Eq. (2.20), of the Differential Riccati Equation (2.14) and of the terminal condition (2.29) is unique within the class of self-adjoint operators $\overline{P}(t)$ such that

$$B^*\overline{P}(t)x \in \begin{cases} C_\gamma([0,T];U) & \text{if } 0 \leq \gamma < \frac{1}{2}, \text{ where } \beta = 0, \quad (2.34) \\ C_{1-\gamma-\varepsilon}([0,T];U) & \text{if } \frac{1}{2} \leq \gamma < 1, \quad (2.35) \end{cases}$$

Theorem 2.3. [F.1], [L-T.22] Under the assumption

$$(-A^*)^\gamma G^*G \in \mathcal{L}(Y), \quad (2.36)$$

which is stronger than assumption (2.27) since $2\gamma-1 < \gamma$, additional regularity results hold true, namely

(i) $$\left|u^0(\cdot,\tau;x)\right|_{C([\tau,T];U)} \leq c_T|x|_Y , \quad x \in Y; \quad (2.37)$$

(ii)   for any $0 \leq \Theta < 1$,

$$\left|(-A^*)^\Theta P(t)\right|_{\mathcal{L}(Y)} \leq \frac{c_T}{1-\Theta} \frac{1}{(T-t)^{\Theta-\gamma}} ; \quad (2.38)$$

(iii) $$B^*P(t) \in \mathcal{L}(Y;C[0,T];U)). \quad \blacksquare \quad (2.39)$$

Remark 2.3. All the above results of Theorem 2.1 are taken from [L-T.22] (or [L-T.4]). They are proved by a variational approach (from the optimal control problem to the Riccati equation) which follows the general strategy of the original contribution [L-T.4], while incorporating an idea of [D-I.1] to quantitatively describe the singularity of the various quantities at $t = T$ via the Banach spaces defined in (2.19). [L-T.4] treated a general second-order parabolic equation defined on a bounded domain Ω of R^n with Dirichlet boundary control, where the constant γ in (1.3) is $\gamma = \frac{3}{4}+\varepsilon$, $\varepsilon > 0$ [T.5], [T.6], [Las.4]. Moreover, in [L-T.4], the operator G was taken to be the identity $G = I$ (with $Z = Y$), certainly a non-smoothing case. The variational approach introduced in [L-T.4] is explicit and constructive in the sense that (1) first, the optimal pair u^0,y^0 is characterized solely in terms of the data of the problem (see (2.3)-(2.5); (ii) next, an operator $P(t)$ is constructed (see (2.20)) in terms of original and optimal evolution, hence ultimately in terms of the original data of the problem; (iii) finally, the operator $P(t)$ is shown to satisfy the Differential Riccati Equation and its limiting condition as $t\uparrow T$. (This constructive and explicit approach is used also for the results of the

subsequent sections for the class (H.2) = (1.6).) The case G = 0 was
previously studied (also for the parabolic problem with Dirichlet
control, and also by abstract methods) in [B.2]. The presence of the
penalization operator G in (1.2) introduces additional genuine
difficulties. Qualitatively, the analyticity of e^{At} tends 'to
compensate' the effects of the unboundedness of B on any interval of
the type [0,T-ε], ∀ ε > 0 small. Instead, the presence of a
non-smoothing operator G produces a singularity at t = T for
$\{L_T^* G^* Ge^{At}x\}(t) = B^* e^{A^*(T-t)} G^* Ge^{At}x$, which occurs in the explicit
formula (2.3) for the optimal $u^0(t,0;x)$. This is reflected by the
quantitative statements of Theorem 2.1: (2.17) for u^0; and (2.18b) for
y^0 when ½ ≤ γ < 1, where the singularity is measured by the spaces
(2.19). This singularity is progressively reduced in Theorem 2.2 under
the smoothing assumption (2.27) on G (vacuous if 0 ≤ γ < ½) and finally
eliminated, see (2.37), if further smoothing is imposed on G as in
(2.36) of Theorem 2.3. Likewise, it is instructive to compare
statements (2.9), (2.32), and (2.39) of increasing regularity for the
gain operator $B^*P(t)$ under progressively stronger smoothing assumptions
on G. The above considerations, in particular (2.18a) and (2.28), show
that for 0 ≤ γ < ½, the optimal y^0 is in C([0,T];Y). This is not
surprising. In fact, standard regularity properties on analytic
semigroup theory yield the well-known result that if γ < ½ in (1.3),
then the operator L in (1.10) is continuous $L_2(0,T;U)) \to C([0,T];Y)$ and
thus *every* solution of (1.1) with $y_0 \in Y$ --not only the optimal
solution y^0 --lies in C([0,T];Y)! Thus, the value γ = ½ gives the
natural 'cutting line' in the range of values of γ, which crucially
bears on the degree of technical difficulties present in the analysis.
The case γ < ½ behaves like the 'B-bounded' case and one has at the
outset the important property that any solution y(t), in particular the
optimal solution $y^0(t,0;y_0)$, belongs to C([0,T];Y). The situation is
more demanding if instead ½ ≤ γ < 1. We have, from (2.3)-(2.5),

$$-u^0(\cdot,\tau;x) = [I_\tau + L_\tau^* R^* RL_\tau]^{-1}\{L_\tau^* R^* Re^{A(\cdot-\tau)}x + L_{\tau T}^* G^* Gy^0(T,\tau;x)\} \quad (2.40)$$

for the optimal control problem on [τ,T], 0 ≤ τ < T, where L_τ, $L_{\tau T}$ are
the operators L in (1.10) and L_T in (2.1) starting now from τ rather
than 0. Crucial to the proof of statements (2.17) for u^0 and (2.18b)
for y^0 is the key property that $[I_\tau + L_\tau^* R^* RL_\tau]^{-1} \in \mathcal{L}(C_\gamma([\tau,T];U))$ with

uniform bound which may be taken independent of τ. This is
accomplished via a boost-strap argument starting from the *a-priori*
L_2-regularity and using the smoothing properties of regularity of the
operators L and L^*. Similarly, crucial to the proof of Theorem 2.2 in
the smoothing case is the key fact that the operator $\Lambda_{\tau T}$ (same as Λ_{0T}
in (2.5) except that the process starts now at τ rather than 0.)
satisfies $\Lambda_{\tau T}^{-1} \in \mathscr{L}(C_{\gamma - \beta}([\tau, T]; U))$ with a uniform bound which may be
taken independent of τ. This is also done by a boost-strap argument.
A boost-strap technique is also behind the proof in [L-T.4] which shows
that the operator $[I + L^* R^* RL]$ is boundedly invertible in the space
$A(\mathscr{F}; U)$ of U-valued functions which are analytic on $\mathscr{F}$ and continuous on
$\bar{\mathscr{F}}$, where $\mathscr{F}$ is an open symmetric set, in the sector of analyticity of
exp(At), based on the interval [0,T]. This step is crucial to obtain
the analyticity properties of the optimal pair ((ix) of Theorem 2.1).
    The regularity properties of $\{u^0, y^0\}$, see (ix) of Theorem 2.1,
as well as those of P(t), see (2.7), (2.8), are *distinctive* of the
analytic class (H.1) = (1.5). They should be contrasted with those
available for the class (H.2) in subsequent sections. A common goal--a
key fact in establishing well-posedness of the Riccati equation--is
that the gain operator $B^* P(t)$ be well defined, which is not *a-priori*
clear when B is unbounded: Eqns. (2.10) in the general case and (2.39)
in the smoothing case are statements of this facts.

Remark 2.4. Another approach, in a sense a converse of the variational
approach described in Remark 2.3, so-called 'direct' (as it proceeds in
reverse from a direct study of the well-posedness of the Riccati
Equation to the optimal control problem via dynamic programming) is
proposed in [F.1], [D-I.1], following [DaP.1]. Here the operator G is
taken to be 'smoothing' for both the purposes of asserting a unique
solution of the Riccati Equation (by local contraction argument and
global *a-priori* bound), as well as for the limiting condition as t↑T.
In these references, a typical smoothing assumption on G are (2.36) for
[Da-I] and (2.27) for [F.1], in which case existence and uniqueness of
the solution to the Differential Riccati Equation is asserted.
Moreover, under (2.27), the various quantities u^0; y^0; $(-A^*)^\Theta P(t)$,
$\Theta \leq \gamma$; $B^* P(t)$ do not experience singularity at t = T any longer as
stated in Theorem 2.3, thereby extending the theory available for
G = 0. In a more recent work [F.5], the direct study of the Riccati
Equation for existence (not for uniqueness) is carried out in the non-

smoothing case for G. Instead of assuming that GL_T is closed

(closable)--a natural hypothesis on G in the variational approach of
[L-T.4], [L-T.22]--[F.5] makes the following assumption on G, say in
the case $G \in \mathcal{L}(Y,W)$:

> there exists a sequence $G_n \in \mathcal{L}(Y,W)$ of operators such that
>
> (a)   there exists $\beta > 2\gamma - 1$ such that each G_n satisfies
>
> the assumption $G_n(-A)^{\beta/2} \in \mathcal{L}(Y,W)$;
>
> (b) $\{G_n^* G_n\}$ is a nondecreasing family of self-adjoint
> operators which converges monotonically to
> $G^* G$ in the sense that as $n \to \infty$:
>
> $$(G_n^* G_n x, x)_Y = \|G_n x\|^2 \uparrow (G^* Gx, x)_Y = \|Gx\|_W^2 ,$$
>
> $$\forall \ x \in Y . \qquad (2.41)$$

Under this assumption (2.41), [F.5, Thm. 3.2] shows existence of a
solution P(t) of the DRE (2.14) (with $\frac{d}{dt}$ $(P(t)x, y)_Y$ on the left side),
which satisfies the regularity properties (2.6) and (2.7), among
others, as in Theorem 2.1 above. In addition, [F.5] obtains also the
strong convergence of P(t) $\to G^* G$ as t↑T (versus weak convergence in
(2.23) of Theorem 2.1), because of the postulated monotonic
approximation property (b). We shall see in Section 2.3 that
assumption (2.41) in [F.5]--which does not invoke B, only A and G--is
stronger than the assumption GL_T closable of Theorem 2.1, which

involves B.

The proof of [F.5, Thm. 3.2] uses an approximating argument
based on two steps reflected by the two properties (a) and (b) of
assumption (2.41):

(i)   The approximating problem involving G_n under the assumed

symmetric condition $(-A^*)^{\beta/2} G_n^* G_n (-A)^{\beta/2} \in \mathcal{L}(Y)$, some $\beta > 2\gamma - 1$,

is handled by the general strategy of a local contraction
principle followed by *a-priori* estimates to yield a global
Riccati solution operator $P_n(t)$ of the DRE (2.14) with endpoint

$G_n^* G_n$ at t = T;

(ii)   the general case under assumption (2.41) is then based on
approximating from below such Riccati solutions $P_n(t)$.

The condition in (a) on $G_n(-A^*)^{\beta/2} \in \mathcal{L}(Y,W)$ is implied by (2.27) [F.1, Lemma 3.1] but does not imply (2.27). Thus the approximating problem in (i) is not fully covered by Theorem 2.2: the proof in [F.5] uses arguments which are new over those in [F.1], [D-I], and which are based on a change of variable introduced in [Da-L-T]: unlike [F.1], [D-I], the more delicate step in [F.5] is now local existence.

2.3.   Counterexamples

It was independently noted in [F.6] and [L-T.22; Remark 5.1] that suitable one-dimensional range (finite range) operators G furnish examples which illustrate the sharpness and/or limitations of the theory presented in Sections 2.1, 2.2.

2.3.1. Counterexample to the existence of the optimal control u^0

The example.   Consider, say the heat equation defined on a (smooth) bounded domain $\Omega \subset R^n$ with $L_2(0,T;L_2(\Gamma))$-control in the Dirichlet boundary conditions, as in Example 6.1, Eq. (6.1), with c = 0 in (6.1a). Here $Y = L_2(\Omega)$, $U = L_2(\Gamma)$. There exists $\phi \in Y$, $|\phi| = 1$ such that

$$\int_0^T |B^* e^{A^*(T-t)} \phi|_U^2 dt = \infty \tag{2.42}$$

for then, otherwise, by transposition, the map $u \to y(T)$ (where $y_0 = 0$) would be continuous $L_2(0,T;L_2(\Gamma)) \to L_2(\Omega) = Y$, which is false even in the one-dimensional case, e.g., [Lio.3; p. 217]. Following [F.6], we consider the associated optimal control problem (1.2) with

$$R = 0; \quad Gy = (y,\phi)_Y\phi; \quad G^* = G = G^*G. \tag{2.43}$$

Note that we have by (2.1) and (2.43),

$$GL_T u = (\int_0^T e^{A(T-t)} Bu(t)dt, \phi)_Y \phi = (u, B^* e^{A^*(T-\cdot)} \phi)_{L_2(0,T;U)} \phi \tag{2.44}$$

so that GL_T is finite rank and unbounded by (2.42), hence *unclosable* [K.1; p. 166].

__Claim__. There is no optimal control in this case. In fact, following [L-T.22], if an optimal control $u^0(\cdot, 0; x) = u^0 \in L_2(0, T; U)$ exists, it satisfies the present version of (2.3), i.e.,

$$-[u^0 + L_T^* G^* G L_T u^0] = L_T^* G^* G e^{AT} x = (e^{AT} x, \phi)_Y B^* e^{A^*(T-t)} \phi, \quad (2.45)$$

where we have used (2.43) on $G^* G$ and (2.2) for L_T^*. Moreover, by (2.43), (2.44),

$$L_T^* G^* G L_T u = L_T^* \{ (L_T u, \phi)_Y \phi \} = (\int_0^T e^{A(T-t)} B u(t) dt, \phi)_Y B^* e^{A^*(T-t)} \phi. \quad (2.46)$$

Using (2.46) in (2.45) yields

$$-u^0 = \{ (u^0, B^* e^{A^*(T-\cdot)} \phi)_{L_2(0, T; U)} + (e^{AT} x, \phi)_Y \} B^* e^{A^*(T-t)} \phi. \quad (2.47)$$

Since $B^* e^{A^*(T-t)} \phi \not\in L_2(0, T; U)$ by (2.42), then (2.47) yields that $u^0 \not\in L_2(0, T; U)$, a contradiction. ∎

__Remark 2.5__. It is argued in [F.6] that, in the present case, it is _not possible_ for the corresponding optimal problem (1.2), (2.43) to satisfy the following three desirable properties:
(i)   that there exists a unique optimal control u^0;
(ii)   that there exists $P(t)$, $0 \leq t \leq T$, non-negative self-adjoint, such that identity (2.13) holds;
(iii) that for every $0 \leq t < T$ and $x \in \mathcal{D}(A)$, $(P(t)x, x)$ is differentiable, $P(t)x \in \mathcal{D}((-A^*)^\gamma)$ and the DRE (2.14) is satisfied. ∎

__Remark 2.6__. We note that the choice (2.42) for ϕ implies that $\phi \not\in \mathcal{D}((-A^*)^{\beta/2})$ for all $\beta > 2\gamma - 1$, and hence [F.5; Sect. 3.1], G in (2.43) does _not_ satisfy assumption (2.41) of [F.5]. In fact, if we had $\phi \in \mathcal{D}((-A^*)^{\beta/2})$ we would obtain that

$$B^* e^{A^*(T-t)} \phi = B^*(-A^*)^{-\gamma}(-A^*)^{\gamma - \beta/2} e^{A^*(T-t)}(-A^*)^{\beta/2} \phi \quad (2.48)$$

would belong to $L_2(0, T; U)$ by (1.3) and analyticity with $2\gamma - \beta < 1$, thus contradicting (2.42). We note that in this case we have $\mathcal{D}((-A^*)^{\beta/2} G^*) = \{0\}$, $\forall \beta > 2\gamma - 1$. ∎

2.3.2. <u>Assumption (2.25) is only sufficient for GL_T to be closed</u>

We shall provide a class of examples where condition (2.25) is *violated*, yet GL_T is *closed*. This is not surprising as condition (2.25)—unlike GL_T—does not involve B. Let the generator A be negative, self-adjoint, say with compact resolvent. (We shall, however, maintain the notation A^*.) Let $\{e_n, n = 1, 2, \cdots\}$ be the corresponding orthonormal basis of eigenvectors of A on Y with eigenvalues $\{-\mu_n\}$, $\mu_n > 0$. Let δ_i, $i = 1, 2$, be two infinite, disjoint sequences of positive integers that exhaust all of the positive integers z: $\delta_1 \cup \delta_2 = z$; $\delta_1 \cap \delta_2 = \emptyset$. Example: $\delta_1 = \{n = 2, 4, 6, \cdots\}$, $\delta_2 = \{n = 1, 3, 5, \cdots\}$. Consider the orthogonal decomposition of Y

$$Y = Y_1 + Y_2, \quad Y_i = \overline{\text{span}} \{e_n, n \in \delta_i\}, \quad i = 1, 2. \quad (2.49)$$

Let Π_i be the orthogonal projection of Y onto Y_i, so that Π_i commutes with A, hence with the semigroup e^{At} and Y_i are invariant under e^{At}. Define a vector $b \in Y_1$ by setting

$$(b, e_n)_Y = \begin{cases} \text{sequence in } n \in \delta_1 \text{ s.t. } \sum_{n \in \delta_1} \mu_n^\beta |(b, e_n)_Y|^2 = \infty \\ 0, \quad n \in \delta_2 \end{cases} \quad (2.50)$$

for all $\beta > 2\gamma - 1$, so that

$$b \notin \mathcal{D}((-A^*)^{\beta/2}), \quad \forall \beta > 2\gamma - 1. \quad (2.51)$$

Next, with $U = Y = W$, define the bounded operators G^*, G and the unbounded operators B^*, B by

$$G^* y = (y_1, a)_Y b + y_2; \quad Gy = (y_1, b)_Y a + y_2; \quad y_i = \Pi_i y \in Y_i; \quad a \in Y. \quad (2.52)$$

$$\begin{cases} By_1 = 0 \\ By_2 = (-A)^\gamma y_2 \end{cases}; \quad \begin{cases} B^* y_1 = 0; \quad y_1 = \Pi_1 y \in Y_1 \\ B^* y_2 = (-A^*)^\gamma y_2; \quad y_2 = \Pi_2 y \in Y_2 \cap \mathcal{D}((-A)^\gamma) \end{cases}. \quad (2.53)$$

One readily obtains by (2.52), (2.51) that

$$\mathcal{D}((-A^*)^{\beta/2}G^*) = \mathcal{D}((-A^*)^{\beta/2}) \cap Y_2;$$

$$(-A^*)^{\beta/2}G^*y = (-A^*)^{\beta/2}y_2, \quad y \in \mathcal{D}((-A^*)^{\beta/2}G^*). \quad (2.54)$$

Thus, $\mathcal{D}((-A^*)^{\beta/2}G^*)$ is *not dense in* Y_2, and condition (2.25) is *violated*.

On the other hand, since $B\Pi_1 u(t) \equiv 0$ and Y_2 is invariant under A and e^{At}, we obtain by (2.1) and (2.52),

$$GL_T u = G\int_0^T e^{A(T-t)}Bu(t)dt$$

$$= G\int_0^T e^{A(T-t)}B\Pi_1 u(t)dt + G\int_0^T e^{A(T-t)}B\Pi_2 u(t)dt$$

$$= G\int_0^T e^{A(T-t)}(-A)^\gamma \Pi_2 u(t)dt = (-A)^\gamma \int_0^T e^{A(T-t)}\Pi_2 u(t)dt, \quad (2.55)$$

where in the last step we have used (2.52) on G, with the integral term in Y_2. Thus, GL_T is a *closed* operator (being the product of a closed, boundedly invertible operator $(-A)^\gamma$ and of a bounded operator [K.1; p. 164]. Our claim is proved. Note that, by (2.2), one likewise has $\{L_T^*G^*y\}(t) = (-A^*)^\gamma e^{A^*(T-t)}y_2, \; Y_2 = \Pi_2 y \in Y_2.$ ∎

2.3.3. <u>Variational versus direct approach: Assumption (2.41) of the direct approach fails, yet</u> GL_T <u>is closed</u>

We have already noted that assumption (2.41) for the direct approach of [F.5; Thm. 3.2] in the case where G is non-smoothing, involves only the operators A and G, not B. Instead, the assumption of the variational approach of Theorem 2.1 that GL_T be closable involves all the data of the problem: G, A, and B. Thus, not surprisingly, we provide new classes of examples where *assumption* (2.41) *fails*, yet GL_T is *closed*. Thus, [F.5; Thm. 3.2] is not applicable, while Theorem 2.1 of [L-T.22] is. We return to the example of Section 2.3.2 and set

$$G_1 y = (y_1, v)_{Y_1} v, \quad G_1^* = G_1 = G_1^* G_1, \quad v \in Y_1, \; |v| = 1, \quad (2.56)$$

where we recall that the subspaces Y_i are invariant under A and e^{At}. It follows from an observation in [F.5; Sect. 3.1] that

G_1 in (2.56) satisfies assumption (2.41)

$$\iff v \in \mathcal{D}((-A^*)^{\beta/2}) \cap Y_1 \quad \text{for some } \beta > 2\gamma-1. \quad (2.57)$$

Next, choose $v = b$, with $b \in Y_1$ the (normalized) vector defined in (2.50), satisfying (2.51). Thus, the operator G_1 ,

$$G_1 y = (y_1, b)_{Y_1} b \text{ does not satisfy (2.41) on } Y_1. \quad (2.58)$$

Since Y_i are invariant under A, it follows that the operator G,

$$G = G_1 + I_2 \text{ does not satisfy (2.41) on Y;} $$

$$G_1 \text{ as in (2.58); } I_2 = \text{identity on } y_2. \quad (2.59)$$

Yet, as seen in Section 2.3.2, GL_T is *closed*. ∎

Remark 2.7. Both approaches, the direct in [F.5] and the variational in [L-T.22] can be extended readily to allow G to be unbounded: $G \in \mathcal{L}(\mathcal{D}(-A)^\rho, Y)$, $\rho > 0$. This will not be pursued here. ∎

Remark 2.8. (i) We have seen in Section 2.3.1 that the operator G in (2.43) with ϕ as in (2.42) neither satisfies assumption (2.41) of the direct approach of [F.5] (see Remark 2.6) nor does it make GL_T closable (see below (2.44)).

(ii) Suppose that $G(-A)^{\beta/2}$ is closable for some $\beta > 2\gamma-1$. Then the assumptions of both approaches are satisfied; i.e., (i) GL_T is closable and (ii) assumption (2.41) holds true. Statement (ii) is proved in [F.5; Section 3.4.1]. Statement (i) follows at once, since then $GL_T = G(-A)^{\beta/2} V_T$ is the product of a closable operator and of a *bounded* operator $V_T u = \int_0^T (-A)^p e^{A(T-t)} (-A)^{-\gamma} Bu(t) dt$, $p = \gamma-\beta/2 < \frac{1}{2}$, so that $V_T \in \mathcal{L}(L_2(0,T;U), Y)$. ∎

3.   **Abstract Differential Riccati Equations for the second class
      subject to the trace regularity assumption (H.2) = (1.6)**

We shall first provide (Section 3.1) the pointwise (in time)
synthesis of the optimal pair under the sole assumption (H.2), with *no*
smoothing required on the observation operator R. Next, under some
minimal assumptions of smoothing on R, further results will be provided
in the specific case of second order hyperbolic equations with control
in the Dirichlet boundary conditions (Section 3.2), including the claim
that the (explicitly constructed) operator P(t) which occurs in the
pointwise synthesis does, in fact, satisfy the DRE. Here the theory is
rather complete for appropriately smooth initial data. Finally, by
imposing an additional requirement of smoothing on R, uniqueness of the
Riccati operator will also be claimed (Section 3.3).

3.1.   **Pointwise synthesis of optimal pair and candidate Riccati
        operator**

Theorem 3.1. [L-T.6, Thm.1.3], [F-L-T, Thm. 2.1] We assume the
regularity hypothesis (H.2) = (1.6) on the dynamics and that, moreover,
$R \in \mathcal{L}(Y;Z)$.   Then:

(i)   there is a unique solution pair of functions $u^0 = u^0(t,0;y_0)$ and
        $y^0 = y^0(t,0;y_0)$, $0 \leq t \leq T$, of the optimal control problem
        (1.1), (1.2) which satisfy

$$u^0 \in L_2(0,T;U), \quad y^0 \in C([0,T];Y); \qquad (3.1)$$

(ii) u^0 and y^0 are related by (recall L, L^* in (1.9),(1.10); L_T, L_T^* in
        (2.1),(2.2))

$$u^0(\cdot,0;y_0) = -L^*R^*R\{y^0(\cdot,0;y_0)\} - L_T^*G^*Gy^0(T,0;y_0) \qquad (3.2)$$

and explicitly given by (see Λ_{0T} in (2.5))

$$\begin{cases} -u^0(t,0;y_0) = \{\Lambda_{0T}^{-1}[L^*R^*R[e^{A\cdot}y_0] + L_T^*G^*Ge^{AT}y_0]\}(t), & (3.3) \\ y^0(t,0;y_0) = \{[I+LL^*R^*R]^{-1}[e^{A\cdot}y_0 - LL_T^*G^*Gy^0(T,0;y_0)]\}(t) & (3.4) \end{cases}$$

$$[I+LL^*R^*R]^{-1} = I - L[I+L^*R^*RL]^{-1}L^*R^*R \in \mathcal{L}(L_2(0,T;Y)); \qquad (3.5)$$

(iii)   there exists an operator $P(t) \in \mathcal{L}(Y)$, given explicitly by

$$P(t)x = \int_t^T e^{A^*(\tau-t)}R^*Ry^0(\tau,t;x)d\tau + e^{A^*(T-t)}G^*Gy^0(T,t;x), \quad x \in Y \quad (3.6a)$$

$$: \text{continuous } Y \to C([0,T];Y); \quad (3.6b)$$

(iv)   (pointwise feedback synthesis)

$$u^0(t,0;y_0) = -B^*P(t)y^0(t,0;y_0) \quad \text{a.e. in } [0,T]; \quad (3.7)$$

(v) $(P(t)x,y)_Y = \int_t^T (Ry^0(\tau,t;x),Ry^0(\tau,t;y))_Z d\tau + (Gy^0(T,t;x),Gy^0(T,t;y))_W$

$$+ \int_t^T (u^0(\tau,t;x),u^0(\tau,t;y))_U d\tau, \quad x,y \in Y; \quad (3.8)$$

(vi) $$P(t) = P^*(t) \geq 0, \quad 0 \leq t \leq T; \quad (3.9)$$

(vii) $$(P(0)x,x)_Y = J(u^0(\cdot,0;x),y^0(\cdot,0;x)). \quad (3.10)$$

(viii) The operator $P(t)$, $0 \leq t < T$, is an isomorphism on Y if and only if the dynamical system (in short the pair $\{A^*,R^*\}$)

$$\eta_t(t) = A^*\eta(t)+R^*g(t), \quad \eta(0) = 0,$$

is exactly controllable on Y over $[0,T-t]$ from the origin with L_2-controls g: i.e., the totality of solution points $\eta(T-t)$ fills all of Y as g runs over all of $L_2(0,T-t;Y)$.

Property (viii) is further pursued in Section 3.4.   ∎

Remark 3.1. Theorem 3.1 is proved in [L-T.6], [F-L-T] by a variational approach. There is no claim in the above generality that the constructively defined operator $P(t)$ is a bonafide solution of the DRE; see more on this in Section 3.4 below. Note that the synthesis property (iv) requires that the operator $B^*P(t)$ be well defined as acting on the optimal trajectory. Instead, the DRE would require that $B^*P(t)$ be well defined on, say, $\mathcal{D}(A)$. Because of the general regularity property (1.12) for the operator L under assumption (H.2), addition to (1.2) of a final state penalization operator G will not cause now essential changes to the analysis, in contrast to the situation of Theorem 2.1 for the class (H.1). In fact, any solution $y(t,\tau;x)$, $x \in Y$ --not only the optimal solution $y^0(t;\tau;x)$ --is continuous in t, $\tau \leq t \leq T$ for τ fixed. This then yields that the map $t \rightarrow y^0(T,t;x)$ is also continuous (using the evolution properties of the operator $x \rightarrow y^0(t,\tau;x)$) and this fact is needed in (3.6a).

3.2. The DRE for second-order hyperbolic equations with Dirichlet control: Existence and properties

In this section a minimal assumption of smoothing is imposed on R, which will then yield that the operator P(t) in (3.6) does satisfy the DRE, in the case of second-order hyperbolic equations with Dirichlet control, of which (7.1) is a canonical example. Thus, we shall specialize the dynamics (1.1) to

$$w_{tt} = -Aw + ADu; \quad \text{i.e., to } A = \begin{vmatrix} 0 & I \\ -A & 0 \end{vmatrix}, \quad Bu = \begin{vmatrix} 0 \\ ADu \end{vmatrix}, \quad (3.11)$$

with $y(t) = [w(t), w_t(t)]$, of the form that arises in mixed problems for second-order hyperbolic equations on a bounded domain $\Omega \subset R^n$, with Dirichlet control, such as (7.1). In (3.11), A is (for simplicity) a positive self-adjoint operator on $X = L_2(\Omega)$ with compact resolvent, D the Dirichlet operator in $\mathcal{L}(U;X)$, $U = L_2(\Gamma)$, defined by (7.4). It is throughout assumed that the regularity hypothesis (H.2) = (1.6) holds true for A and B as in (3.11) on the space

$$Y \equiv X \times [\mathcal{D}(A^{\frac{1}{2}})]' = L_2(\Omega) \times H^{-1}(\Omega), \quad (3.12)$$

where duality is with respect to the X-topology; and moreover, that

$$R = \begin{vmatrix} R_1 & 0 \\ 0 & R_2 \end{vmatrix} \in \mathcal{L}(Y;Z). \quad (3.13)$$

Thus, Theorem 3.1 holds true. Moreover,

Theorem 3.2. [L-T.6] (a) Assume, in addition, that

$$R_1^* R_1: \text{ continuous } H^{\frac{1}{2}-2\delta}(\Omega) = \mathcal{D}(A^{\frac{1}{4}-\delta}) \rightarrow \mathcal{D}(A^{\frac{1}{4}}), \quad (3.14a)$$

$$R_2^* R_2: \text{ continuous } H^{-\frac{1}{2}-2\delta}(\Omega) = [\mathcal{D}(A^{\frac{1}{4}+\delta})]' \rightarrow [\mathcal{D}(A^{\frac{1}{4}})]', \quad (3.14b)$$

where R_1^* is the X-adjoint of R_1 and R_2^* is the $[\mathcal{D}(A^{\frac{1}{2}})]'$-adjoint of R_2, so that (3.14a-b) collectively mean that $R^* R$: continuous

$Y_r \to \mathcal{D}(A^{\frac{1}{4}}) \times [\mathcal{D}(A^{\frac{1}{4}})]'$, where R^* is the Y-adjoint of R and where the space (of regular data) Y_r is defined by

$$Y_r \equiv \mathcal{D}(A^{\frac{1}{4}-\delta}) \times [\mathcal{D}(A^{\frac{1}{4}+\delta})]' = H^{\frac{1}{2}-2\delta}(\Omega) \times H^{-\frac{1}{2}-2\delta}(\Omega). \qquad (3.15)$$

Then, the following regularity holds true for the optimal pair $\{u^0, y^0\}$, $y^0[w^0, w_t^0]$ for initial data $y_0 = [w_0, w_1] \in Y_r$:

$$u^0 \in H^{\frac{1}{2}, \frac{1}{2}}(\Sigma), \text{ a fortiori } u^0 \in C([0,T]; L_2(\Gamma)); \qquad (3.16)$$

$$w^0 \in C([0,T]; \mathcal{D}(A^{\frac{1}{4}-\delta})) \cap H^{\frac{1}{2}-2\delta}(0,T;X); \qquad (3.17)$$

$$w_t^0 \in C([0,T]; [\mathcal{D}(A^{\frac{1}{4}+\delta})]'. \qquad (3.18)$$

(b) Let now $\{R_1, R_2\}$ satisfy, in addition to (3.13),

$$R_1^* R_1: \text{ continuous } H^{\frac{1}{2}-2\delta}(\Omega) = \mathcal{D}(A^{\frac{1}{4}-\delta}) \to \mathcal{D}(A^{\frac{1}{4}+\delta}) = H_0^{\frac{1}{2}+2\delta}(\Omega); \qquad (3.19a)$$

$$R_2^* R_2: \text{ continuous } H^{-\frac{1}{2}-2\delta}(\Omega) = [\mathcal{D}(A^{\frac{1}{4}+\delta})]' \to [\mathcal{D}(A^{\frac{1}{4}-\delta})]' \qquad (3.19b)$$

(slightly more restrictive than (3.14)). Then

(b_1)
$$B^* P(t) \in \mathcal{L}(Y_r; C([0,T]; L_2(\Gamma)). \qquad (3.20)$$

(b_2) (Existence) The non-negative self-adjoint operator P(t) defined by (3.6a) satisfies the DRE for all $x, z \in Y_r$, and $0 \le t < T$,

$$\begin{cases} \dfrac{d}{dt}(P(t)x,z)_Y = -(Rx,Rz)_Z - (P(t)x,Az)_Y - (P(t)Ax,z)_Y + (B^*P(t)x, B^*P(t)z)_U ; \\ P(T) = G^*G. \end{cases}$$
$$\qquad (3.21)$$

Remark 3.2. The above result was originally proved in [L-T.6] in the case of the wave equation (or second-order hyperbolic equations) with Dirichlet control, by a combination of abstract methods and p.d.e. methods (once the regularity property (H.2) = (1.6) has been ascertained [L-T.2], [Lio.1], [L-L-T]. In this case we have (with equivalent norms)

$$X = L_2(\Omega); \quad \mathcal{D}(A^{\frac{1}{2}}) = H_0^1(\Omega); \quad [\mathcal{D}(A^{\frac{1}{2}})]' = H^{-1}(\Omega); \qquad (3.22)$$

$$\mathcal{D}(A^{\frac{1}{4}-\delta}) = H^{\frac{1}{2}-2\delta}(\Omega); \quad \mathcal{D}(A^{\frac{1}{4}+\delta}) = H_0^{\frac{1}{2}+2\delta}(\Omega); \quad [\mathcal{D}(A^{\frac{1}{4}+\delta})]' = H^{-\frac{1}{2}-2\delta}(\Omega);$$
(3.23)

$$Y = L_2(\Omega) \times H^{-1}(\Omega); \quad Y_r = H^{\frac{1}{2}-2\delta}(\Omega) \times H^{-\frac{1}{2}-2\delta}(\Omega).$$
(3.24)

A key issue in the proof of (3.16)-(3.18) is that $[I_r + L_r L_r^* R^* R]^{-1} \in \mathcal{L}(Y; L_2(0,T;Y_r))$ with uniform bound which may be taken independent of r. This is achieved by compactness arguments. A companion paper for first-order hyperbolic systems, see Section 7.4 below, is [Ch-L]. The combination of abstract and p.d.e. methods of these papers should be extendable to, say, some fourth-order operators A. ∎

3.3. DRE: Existence and uniqueness

Imposing a stronger assumption of smoothing on the operator R yields also uniqueness of the Riccati operator. In addition to (H.2) = (1.6) on the dynamics and R ∈ L(Y;Z) on the observation, we may assume in this subsection:

(A.1): the map $R^* R e^{At} B$ can be extended as a map: continuous
$U \to L_1(0,T;Y)$:

$$\int_0^T \|R^* R e^{At} Bu\|_Y dt \le c_T \|u\|_U.$$
(3.25)

(A.2): the map $B^* e^{A^* t} G^*$ can be extended as a map continuous
$Y \to L_\infty(0,T;U)$:

$$\sup_{0 \le t \le T} |B^* e^{A^* t} G^* x|_U \le c_T |x|_Y, \quad x \in Y.$$
(3.26)

Theorem 3.3. [DaP-L-T.1] Under the above assumptions, there exists a non-negative self-adjoint operator $P(t) = P^*(t) \ge 0$, $0 \le t \le T$, given explicitly by the same formula (3.6a) (where y^0 is the optimal trajectory guaranteed by Theorem 3.1(i)) such that

(i) $\qquad\qquad P(t) \in \mathcal{L}(Y; C([0,T];Y);$
(3.27a)

(ii) $\qquad\qquad B^* P(t) \in \mathcal{L}(Y; C([0,T];U);$
(3.27b)

(iii) $\quad B^* P(t) e^{A(t-r)} B \in \mathcal{L}(U; L_2(r,T;U)$ uniformly in r,

$$\sup_{0 \leq \tau \leq T} \int_{\tau}^{T} \|B^* P(t) e^{A(t-\tau)} Bu\|_U^2 \, dt \leq c_T \|u\|_U^2 \ . \qquad (3.28)$$

(iv)   The unique optimal pair $\{u^0, y^0\}$ satisfies the pointwise feedback synthesis property (3.7) (except that it is now for all $t \in [0,T]$) as well as properties (v) = (3.8) and (vii) = (3.10) for the optimal cost of Theorem 3.1.

(v)   (Existence) For $0 \leq t < T$, the operator $P(t)$ satisfies the DRE (3.21), now for all $x, z \in \mathcal{D}(A)$, as well as the terminal condition $P(T) = G^* G$.

(vi)   (Uniqueness) The operator $P(t)$ given by formula (3.6a) is the unique solution of the DRE as in point (v) above, within the class of non-negative self-adjoint operators which satisfy properties (i) = (3.26), (ii) = (3.27), (iii) = (3.28). ∎

Remark 3.3. The above result was proved in [DaP-L-T] by a 'direct' method (from the DRE to the optimal control problem): this first establishes well-posedness of the DRE (3.21) and next constructs, via dynamic programming, the optimal control problem which generates the original DRE. Well-posedness of the DRE (3.21) (for all $x, y \in \mathcal{D}(A)$) is achieved, following the original strategy in [DaP.] for the "B-bounded" case, by a local contraction argument near T, followed by global *a-priori* bounds. This strategy encounters additional technical difficulties to be sure. In particular, a new change of variable is used in [D-L-T]. This way (existence and) uniqueness (in the sense of Theorem 3.3(vi)) is obtained, at the price of the smoothing assumption (A.1) = (3.25) on R, a quantitative statement thereof will be given in Remark 3.4 in the case of the wave equation with Dirichlet control and in Section 7.4 in the case of first-order hyperbolic systems.

Instead, the prior work [L-T.6], [C-L.1] followed a variational approach (from the optimal control problem to the DRE) leading to Theorem 3.2 which has a markedly weaker assumption of smoothing on R, but does not claim uniqueness. ∎

Remark 3.4. (Wave equation with Dirichlet control) We now return to (3.11), the wave equation (or a second-order hyperbolic equation) with Dirichlet control, to be analyzed in more details in Section 7.1 below. The relevant spaces are given by (3.22)-(3.24) of Remark 3.2. Plainly the regularity assumptions (3.14), or (3.19), of Theorem 3.2 hold true if $R_1^* R_1$ has a "smoothing action of the order of $A^{-\varepsilon}$" (technically, say (3.19b) is equivalent to $A^{\frac{1}{4}+\delta} R_1^* R_1 A^{-\frac{1}{4}+\delta} \in \mathcal{L}(X)$, $X = L_2(\Omega)$ in our case). In contrast, it can be shown [Da-L-T] that Theorem 3.3 requires for its assumption (A.1) = (3.25) to be satisfied that

$$R_1^* R_1 A^{\frac{1}{4}+\varepsilon} \in \mathcal{L}(L_2(\Omega)); \quad R_2^* R_2 A^{\frac{1}{4}+\varepsilon} \in \mathcal{L}(H^{-1}(\Omega)). \qquad (3.29)$$

3.4. Non-smoothing case: Weaker notions of solution

For the class (H.2) = (1.6) of dynamics, under the sole assumption that $R \in \mathcal{L}(Y,Z)$, Theorem 3.1 provides the non-negative self-adjoint operator $P(t)$ in (3.6) needed for the pointwise synthesis (3.7), as well as several of its properties. What is conspicuously absent in the statement of Theorem 3.1 is a claim that in this generality such $P(t)$ is a solution of the DRE. No such claim is available at present, the most general statements of existence being the ones of Theorem 3.2 for second-order hyperbolic equations with Dirichlet control [L-T.6], the results of [Ch-L.1] for first-order hyperbolic systems (see Section 7.4), and Theorem 3.3 [Da-L-T] (see also Theorem 4.1 which follows). When R is only in $\mathcal{L}(Y;Z)$, lack of (proof of) regularity properties of the gain operator $B^* P(t)$ prevents one from justifying the formal steps leading to the desired conclusion that such operator $P(t)$ satisfies the DRE (3.21) for, say, $x, y \in \mathcal{D}(A)$. Note that, at least when the pair $\{A^*, R^*\}$ is exactly controllable, say, on $[0,T]$, the operator $P(t)$, $0 \leq t < T$, is an isomorphism on Y (Thm. 3.1(viii)) and hence $B^* P(t)$ is bounded from Y to U if and only if so is B, the trivial case. Thus, in general, $B^* P(t)$ may be unbounded, and it is an issue whether e.g. is even densely defined.

Under these circumstances, it is of interest to regard the constructed operator $P(t)$ of Theorem 3.1 as "solution" of the corresponding Riccati Differential Equation (3.21) in a suitably weaker sense.

Viscosity solution. One approach to this may be given by seeing such $P(t)$ as limit of appropriate Riccati operators $P_\varepsilon(t)$ of regularizing problems, where all $P_\varepsilon(t)$ satisfy the DRE (3.21).

Regularizing problems. We introduce a parameter of regularization $\varepsilon_0 \geq \varepsilon > 0$, $\varepsilon \downarrow 0$, and consider the family $\{R_\varepsilon\}$ of observation operators satisfying

$$R_\varepsilon \in \mathcal{L}(Y;Z); \quad R_\varepsilon \to R \text{ strongly: } R_\varepsilon x \to Rx, \quad x \in Y. \qquad (3.30)$$

$$(A.1_\varepsilon) \qquad \int_0^T \|R_\varepsilon^* R_\varepsilon e^{At} Bu\|_U dt \le c_{T,\varepsilon} \|u\|_U. \qquad (3.31)$$

Remark 3.5. Such family always exists and we may in fact take $R_\varepsilon = R \frac{1}{\varepsilon} R(\frac{1}{\varepsilon}, A)$ in terms of the resolvent of A, so that

$$\|R_\varepsilon^* R_\varepsilon e^{At} Bu\|_U = \frac{1}{\varepsilon^2} \|R(\frac{1}{\varepsilon}, A^*) R^* R e^{At} R(\frac{1}{\varepsilon}, A) Bu\|_U \le \frac{1}{\varepsilon^2} c_{T,\varepsilon} \|u\|_U,$$

since $\|R(\frac{1}{\varepsilon}, A)B\| \le c_\varepsilon$ by (1.3) with $\gamma = 1$, and (3.31) follows at once. ∎

With each R_ε we associate the corresponding optimal control problem (OCP)$_\varepsilon$ on $[0,T]$, $T < \infty$, yielding the unique pair of optimal solutions $\{u_\varepsilon^0(t,0;y_0), y_\varepsilon^0(t,0;y_0)\}$ and the corresponding operator $P_\varepsilon(t)$ by, say, Theorem 3.1 or Theorem 3.3. Moreover, by Theorem 3.3, $P_\varepsilon(t)$ is the *unique solution of the corresponding* DRE$_\varepsilon$ as explained there. The desired connection between the original $P(t)$ and the constructed regularizing family of non-negative self-adjoint bona fide Riccati operators $P_\varepsilon(t)$ is the following result, which was originally envisioned for numerical purposes. It makes the operator $P(t)$ of (3.6) in Theorem 3.1 a "viscosity" solution of the DRE (3.21).

Theorem 3.4. [Las.5] We have the following strong limits as $\varepsilon \downarrow 0$:

(i) $\qquad P(t)x = \lim P_\varepsilon(t)x, \quad x \in Y; \qquad (3.32)$

(ii) $\qquad u^0(\cdot,0;y_0) = \lim u_\varepsilon^0(\cdot,0;y_0) \quad \text{in } L_2(0,T;Y); \qquad (3.33)$

(iii) $\qquad y^0(\cdot,0;y_0) = \lim y_\varepsilon^0(\cdot,0;y_0) \quad \text{in } C([0,T];Y); \qquad (3.34)$

(iv) $\qquad J(u^0,y^0) = \lim J(u_\varepsilon^0,y_\varepsilon^0). \quad ∎ \qquad (3.35)$

The Dual Differential Riccati Equation (DDRA). Under the circumstances of property (viii) of Theorem 3.1 we have seen that the (candidate) Riccati operator $P(t)$ given by (3.6) is an isomorphism on Y for each $t \in [0,T)$ if and only if the pair $\{A^*, R^*\}$ is exactly controllable on $[0,T]$. Thus, in this case, we let $Q(t)$ be the non-negative self-adjoint operator on Y:

$$Q(t) = P^{-1}(t), \quad 0 \leq t < T, \tag{3.36}$$

and we readily verify that $Q(t)$ formally satisfies the following Dual Differential Riccati Equation for all $x, z \in \mathcal{D}(A^*)$ and $0 \leq t < T$:

$$\frac{d}{dt} (Q(t)x, z)_Y = -(B^*x, B^*z)_U + (Q(t)x, A^*z)_Y$$

$$+ (Q(t)A^*x, z)_Y + (RQ(t)x, RQ(t)z)_Z. \tag{3.37}$$

The DDRE (3.37) arises from a quadratic cost optimal control problem, which however requires that A be a s.c. *group* generator; see Section 5, Corollary 5.7 and ff. for more details in the case $T = \infty$.

Following [F.3] one may introduce a weaker notion of solution of the original DRE (2.14) as follows.

(i) One first *assumes* that the end point condition $P(T) = P_T$ of the DRE (2.14) be a non-negative self-adjoint *isomorphism* on Y and studies the DDRE (2.37) in $Q(t)$ with end-condition $Q(T) = Q_T = P_T^{-1}$. Note that the DDRE (2.37) has the quadratic term involving the bounded operator R, while the original DRE (2.14) in $P(t)$ has the quadratic term involving the unbounded operator B^*. Thus the DDRE (2.37) is easier to handle than the DRE (2.14). In fact, the general strategy of local contraction coupled by global *a-priori* bounds [DaP.1] readily yields a unique solution $Q(t)$ of this dual problem [F.3]. Next, [F.3] proves that $Q(t)$ is an isomorphism $0 \leq t \leq T$. One can thus define the operator $P(t)$ by

$$P(t) = Q^{-1}(t), \quad 0 \leq t \leq T. \tag{3.38}$$

Then [F.3] proves that such $P(t)$ satisfies the identities (3.6a) and (3.8) of Theorem 3.1. There is *no claim*, however, *that* $P(t)$ *defined by* (2.38) *solves* the *original* DRE (2.14).

(ii) The general case where P_T is not an isomorphism is reduced step (i) by an approximation argument. More precisely, [F.3]

approximates the general P_T by $P_T+\varepsilon I$, $\varepsilon \downarrow 0$, which is an isomorphism.
Step (i) gives then operators $P_\varepsilon(t)$ defined by $P_\varepsilon(t) = Q_\varepsilon^{-1}(t)$
satisfying identities (3.6a) and (3.8). By letting $\varepsilon \downarrow 0$, [F.3] obtains
an operator $P(t) = \lim P_\varepsilon(t)$ in $\mathcal{L}(Y)$ uniformly in $[0,T]$, which likewise
satisfies identities (3.6a) and (3.8). Again there is *no claim that*
such $P(t)$ *solves the original* DRE (2.14), *nor that such limit* $P(t)$ *is*
an isomorphism. Nevertheless, the operator $P(t)$ constructed in this
fashion may be viewed as a "weak notion" of solution of the original
DRE (2.14).

4.   **Abstract Differential Riccati Equations for the second class**
        subject to the regularity assumptions $(H.2_R) = (1.8)$

Orientation. In the present section we shall adopt the variation
$(H.2_R) = (1.8)$ of the regularity assumption $(H.2) = (1.6)$. In
addition, we shall allow the observation operator R to be unbounded.
Thus, in the present setting, the unbounded coefficient operators A and
B give rise to a (possibly) *unbounded* input-solution operator L in
(1.10) from $L_2(0,T;U)$ to $L_2(0,T;Y)$ and, moreover, the observation
operator R is likewise (possibly) *unbounded* from Y to Z. This setting
is not covered by the one of Sections 3.1, 3.3 as $(H.2) = (1.6)$ assumed
there amounts to L continuous $L_2(0,T;U) \to C([0,T];Y)$, see (1.12). Yet
there are important and natural boundary control problems--such as the
Neumann boundary control/Dirichlet boundary observation problem for
second-order hyperbolic equations described in the subsequent Section
8--which give rise precisely to this situation. It is, in fact, the
desire to cover this hyperbolic problem that motivates the present
section and provides a guiding example. Technically, the setting of
Section 3.3 is not included in the present setting either, strictly
speaking, unless R is an isomorphism; see Remark 1.1. (However, we may
say that 'morally' the present level of generality includes the setting
of Section 3.3, in the sense that the technique of proof of [L-T.10] of
the following Theorem 4.1 of this section would also apply to the
setting of Section 3.3 and produce Theorem 3.3). As already remarked,
and in line with the spirit of this paper, the present abstract
framework is motivated by the 'concrete' quadratic cost problem for the
wave equation with $L_2(\Sigma)$-Neumann control, which penalizes also the

$L_2(\Sigma)$-norm of the (Dirichlet) *trace* of its solutions. In this example, the observation operator R is the Dirichlet trace, and it happens that the composition RL--which gives the maps from the control space $L_2(0,T;L_2(\Gamma))$ to the space $L_2(0,T;L_2(\Gamma))$ of the Dirichlet observations of the hyperbolic solutions--is, in fact, *bounded*; i.e., it is nicer than each of its components viewed separately. This says that the (Dirichlet) trace of the solution behaves more regularly than it could be inferred from looking at the interior regularity of the solution and applying trace theory (even formally). This property is, in fact, a distinctive feature of waves and plates problems, as it has been discovered in recent years in a variety of situations.

Section 4.1 will provide the major theoretical results and Section 8 will be devoted to the boundary control/boundary observation illustration. We must hasten to point out that verification that all required abstract assumptions are satisfied in the case of the wave equation problem with boundary observation is not a trivial task: it requires, in a *critical* way, the sharp regularity theory [L-T.20], [L-T.23] that has become available only very recently, while earlier regularity results [L-M], [M.1], for second-order hyperbolic mixed problems of Neumann type is inadequate and insufficient. This will be seen more technically in Section 8.1.

4.1. Theoretical results: Theorems 4.1 and 4.2

The following abstract assumptions (as (1.3) and $(H.2_R)$) will either capture, or else properly contain, intrinsic *properties* of the hyperbolic problem of Section 4.2.

Assumptions. In addition to the standing assumptions (1.3) and $(H.2_R) = (1.8)$, we shall assume that $G = 0$ and that

(h.0) $R \in \mathcal{L}(\mathcal{D}(A);Z)$;   equivalently $RA^{-1} \in \mathcal{L}(Y;Z)$ (4.1)

   ($\mathcal{D}(A)$ endowed with norm $\|y\|_{\mathcal{D}(A)} = \|Ay\|_Y$, equivalent to the graph norm, as A is assumed boundedly invertible without loss of generality);

(h.1)   the map $Re^{At}B$ can be extended as a map: continuous
   $U \to L_1(0,T;Z)$:

$$\int_0^T \|Re^{At}Bu\|_Z dt \leq c_T \|u\|_U , \quad u \in U; \tag{4.2}$$

(h.2) the map Re^{At} can be extended as a map: continuous $Y \to L_\infty(0,T;Z)$:

$$\sup_{0 \leq t \leq T} \|Re^{At}x\|_Z \leq c_T \|x\|_Y , \quad x \in Y. \tag{4.3}$$

<u>Theorem 4.1</u>. [L-T.10] Under the above assumptions (h.0) = (4.1) through (h.2) = (4.3) (in addition to (1.3) and (H.2$_R$) = (1.8)), there exists a unique solution $\{u^0(t,0;y_0), y^0(t,0;y_0)\}$ of problem (1.1), (1.2), given explicitly by (see (1.9), (1.10)),

$$-u^0(t,0;y_0) = \{[I+L^*R^*RL]^{-1}[L^*R^*(Re^{A\cdot}y_0)]\}(t). \tag{4.4}$$

$$y^0(t,0;y_0) = e^{At}y_0 + (Lu^0)(t); \tag{4.5a}$$

$$Ry^0(t,0;y_0) = \{[I-RL(I+L^*R^*RL)^{-1}L^*R^*][Re^{A\cdot}y_0]\}(t). \tag{4.5b}$$

Moreover, there exists a solution $P(t) \in \mathcal{L}(Y)$, $0 \leq t \leq T$ of the DRE

$$\begin{cases} \frac{d}{dt}(P(t)x,z)_Y = -(Rx,Rz)_Z - (P(t)Ax,z)_Y - (P(t)x,Az)_Y + (B^*P(t)x, B^*P(t)z)_U , \\ \qquad\qquad\qquad\qquad\qquad\qquad\qquad\qquad\qquad \forall\ x,z \in \mathcal{D}(A), \\ P(T) = 0, \end{cases} \tag{4.6}$$

which is, in fact, given constructively by

$$P(t)x = \int_t^T e^{A^*(\tau-t)}R^*Ry^0(\tau,t;x)d\tau, \quad x \in Y. \tag{4.7}$$

Such $P(t)$ enjoys the following properties:

(i) $\qquad\qquad\qquad P(t) = P^*(t), \quad 0 \leq t \leq T; \tag{4.8}$

(ii) $\qquad\qquad\qquad P(t) \in \mathcal{L}(Y;C([0,T];Y)); \tag{4.9}$

(iii) $\qquad\qquad\qquad B^*P(t) \in \mathcal{L}(Y;C([0,T];U)); \tag{4.10}$

(iv) $\quad B^*P(t)e^{A(t-\tau)}B \in \mathcal{L}(U;L_2(\tau,T;U))$, uniformly in τ:

$$\sup_{0 \leq \tau < T} \int_{\tau}^{T} \| B^* P(t) e^{A(t-\tau)} Bu \|_U^2 dt \leq c_T \| u \|_U^2 \; ; \qquad (4.11)$$

(v)   (feedback synthesis) the optimal pair is related by

$$u^0(t,0;y_0) = -B^* P(t) y^0(t,0;y_0), \quad 0 \leq t \leq T; \qquad (4.12)$$

(vi) $(P(t)x,z)_Y = \int_t^T (Ry^0(\tau,t;x), Ry^0(\tau,t;z))_Z d\tau$

$$+ \int_t^T (B^* P(\tau) y^0(\tau,t;x), B^* P(\tau) y^0(\tau,t;z))_U d\tau,$$

from which the optimal cost is

$$J(u^0(\cdot,\tau;y_0), y^0(\cdot,t;y_0)) = (P(\tau)y_0, y_0)_Y , \qquad y_0 \in Y. \quad (4.13)$$

(vii) (Uniqueness) The operator $P(t)$ in (4.7) is the unique solution
to enjoy properties (i) = (4.8) through (iv) = (4.11). ∎

A more regular case is given by the following result. To
state it, we shall postulate the existence of a Hilbert space
$\mathcal{U}_{[0,T]} \subset L_2(0,T;U)$ (algebraically and topologically) with restriction
$\mathcal{U}_{[t,T]} \subset L_2(t,T;U)$, such that the following assumptions, (h.3) through
(h.5), hold true where (see (1.19)):

$$(L_t u)(\tau) = \int_t^\tau e^{A(\tau-r)} Bu(r) dr. \qquad (4.14)$$

(h.3) L_0: continuous $\mathcal{U}_{[0,T]} \to C([0,T];Y);$ $\qquad (4.15)$

(h.4)   the map $L_0^* R^* [Re^{A \cdot}]$ can be extended as a map:

$$L_0^* R^* [Re^{A \cdot}]: \text{continuous } Y \to \mathcal{U}_{[0,T]}; \qquad (4.16)$$

(h.5)   for each $t \in [0,T]$, $T < \infty$, the map

$$L_t^* R^* R L_t \text{ is compact } \mathcal{U}_{[t,T]} \to \text{itself.} \qquad (4.17)$$

Remark 4.1. Assumption (h.4) = (4.16) is stronger than assumptions (h.2) = (4.3) and (H.2$_R$) = (1.8) combined [L-T.10, Remark 6.2]. ∎

The next theorem gives regularity results for the optimal pair $\{u^0(\cdot,\tau;y_0),y^0(\cdot,\tau;y_0)\}$.

Theorem 4.2. [L-T.10] (i) Assume hypotheses (h.4) = (4.16) and (h.5) = (4.17). Then

$$\sup_{0\leq\tau\leq T}\|u^0(\cdot,\tau;x)\|_{\mathcal{U}_{[t,T]}} \leq c_T\|x\|_Y . \qquad (4.18)$$

(ii) Assume, in addition, hypothesis (h.3) = (4.15). Then

$$\sup_{0\leq\tau\leq T}\|y^0(\cdot,\tau;x)\|_{C([\tau,T];Y)} \leq c_T\|x\|_Y . \quad ∎ \qquad (4.19)$$

Application of these two theorems will be given in Section 8.

5.   **Abstract Algebraic Riccati Equations: Existence and uniqueness**

In this section where $T = \infty$, we shall treat the general situation where the s.c. semigroup exp(At) is generally unstable on Y, i.e., with $\omega_0 = \lim[(\ln\|\exp(At)\|)/t] > 0$ as $t \to +\infty$ in the uniform norm $\mathcal{L}(Y)$, so that $\|e^{At}\| \leq Me^{(\omega_0+\varepsilon)t}$, $\forall\ \varepsilon > 0$, $t \geq 0$, and M depending on $\omega_0+\varepsilon$. We then consider throughout the translation $\hat{A} = -A+\omega I$, ω = fixed > ω_0, so that $\hat{A}$ has well-defined fractional powers on Y and $-\hat{A}$ is the generator of an s.c. semigroup $e^{-\hat{A}t}$ on Y satisfying $\|e^{-\hat{A}t}\| \leq \hat{M}e^{-\hat{\omega}t}$, $t \geq 0$; $\hat{\omega} = \omega-\omega_0-\varepsilon > 0$. Moreover, G = 0 in (1.2) while R is non-smoothing

$$R \in \mathcal{L}(Y;Z) . \qquad (5.0)$$

In this section we shall then discuss the solvability of the following abstract Algebraic Riccati Equation (ARE)

$$(Px,Ay)_Y+(PAx,y)_Y+(Rx,Ry)_Z-(B^*Px,B^*Py)_U = 0;$$

$$\forall\ x,y \in \mathcal{D}(A). \qquad (5.1)$$

As is well known, a solution P of this equation (if it exists and it possesses certain regularity properties) provides the sought-after feedback operator which occurs in the representation (1.18) of the optimal control law. In the case where B is an unbounded operator, the obvious difficulty is related to the interpretation of the 'gain operator' B^*P, which *a priori* need not be even densely defined on Y (even when the existence of a "Riccati operator" $P \in \mathcal{L}(Y)$ is asserted). The crux of the matter is therefore this: To prove that the Riccati operator P possesses certain 'regularity' properties which will guarantee a proper definition of the gain operator B^*P, at least as an unbounded operator with dense domain in Y for the representation (1.18) to be meaningful. We begin with the first class ('analytic') which has received more attention in the literature.

5.1. Algebraic Riccati Equation for the first class subject to the analyticity assumption (H.1) = (1.5)

Theorem 2.1 ([D-I], [F.2], [L-T.7]).

I. **Existence**. For the first class covered by hypothesis (H.1) = (1.5) and subject to the Finite Cost Condition (1.9), there exists a self-adjoint, non-negative definite solution $0 \leq P = P^* \in \mathcal{L}(Y)$ of the ARE (5.1) such that:

(i) $(\hat{A}^*)^{1-\epsilon}P \in \mathcal{L}(Y)$, $\forall \epsilon > 0$;   (5.2)

and indeed ϵ may be taken $\epsilon = 0$ if the original A is self-adjoint or normal or has a Riesz basis of eigenvectors on Y; thus, P is compact if A has compact resolvent;

(ii) $B^*P \in \mathcal{L}(Y,U)$;   (5.3)

(iii) $J(u^0,y^0) = (Py_0,y_0)_Y$;   (5.4)

(iv) $u^0(t;y_0) = -B^*Py^0(t)y^0(t;y_0)$ for all $0 < t < \infty$.   (5.5)

II. **Regularity of the optimal pair**. For each fixed $y_0 \in Y$, the functions $y^0(t;y_0)$ and $u^0(t;y_0)$ are analytic in t as Y-valued or U-valued functions, a consequence of the analyticity of the feedback semigroup $e^{A_P t}$ below in (5.11) and of (5.3).

Remark. In the case where (1.1) models a second order parabolic equation on a bounded domain $\Omega \subset R^n$ with Dirichlet boundary control, the following additional regularity properties of the optimal pair hold true [L-T.7]:

$$\text{if } y_0 \in L_2(\Omega) \Rightarrow \begin{cases} e^{-\omega t}y^0(\cdot\,;y_0) \in H^{1-2\varepsilon,\frac{1}{2}-\varepsilon}(Q_\infty), \\ \qquad\qquad\qquad Q_\infty = (0,\infty)\times\Omega, \qquad (5.6) \\ e^{-\omega t}u^0(\cdot\,;y_0) \in H^{\frac{1}{2}-2\varepsilon',\frac{1}{4}-\varepsilon'}(\Sigma_\infty), \\ \quad \forall\, \varepsilon' > \varepsilon > 0,\ \Sigma_\infty = (0,\infty)\times\Gamma; \qquad (5.7) \end{cases}$$

$$\text{if } y_0 \in H^{\frac{1}{2}-\rho}(\Omega) \Rightarrow \begin{cases} e^{-\omega t}y^0(\cdot\,;y_0) \in H^{\frac{1}{2}-2\rho,\frac{1}{4}-\rho}(Q_\infty), \\ \qquad\qquad\qquad\qquad \rho > 0, \qquad (5.8) \\ e^{-\omega t}u^0(\cdot\,;y_0) \in H^{2-2\rho',1-\rho'}(\Sigma_\infty), \\ \qquad\qquad\qquad\qquad \rho' > \rho > 0. \qquad (5.9) \end{cases}$$

III.   **Uniqueness**. In addition to the assumption of part I, we assume that the following so-called 'detectability condition' (D.C.) holds:

(D.C): $\begin{cases} \text{There exists } K \in \mathcal{L}(Z,Y) \text{ such that the} \\ \text{s.c. semigroup } e^{(A+KR)t} \text{ generated by } A+KR \\ \text{is exponentially (uniformly) stable on } Y. \end{cases}$ (5.10)

Then

(a)   the solution P to the ARE (5.1) is unique within the class of non-negative self-adjoint operators in $\mathcal{L}(Y)$, which satisfy the regularity requirement (5.3);

(b)   the s.c., analytic semigroup $e^{A_P t}$ generated by $A_P = A-BB^*P$ is exponentially (uniformly) stable on Y:

$$\|e^{A_P t}\|_{\mathcal{L}(Y)} \le M_P e^{-\omega_P t}, \quad t > 0 \qquad (5.11)$$

for some constants M_P, $\omega_P > 0$. ∎

Remark 5.0. If the original semigroup exp(At) is (uniformly) stable, i.e., $\omega_0 < 0$ for the constant above (5.0), then one can give an explicit, constructive formula for P in terms of the optimal dynamics, which in turn is given explicitly in terms of the data of the problem; precisely as in the case $T < \infty$, seen before in (2.20). If instead, $\omega_0 > 0$, then the explicit formula for P becomes actually an identity satisfied by P; see e.g., [L-T.7; Section 2]. ∎

As in the case $T < \infty$, two distinct, yet complementary,
approaches are available to prove Theorem 5.1 (existence and
uniqueness): (i) a variational approach [L-T.7], [L-T.19], and (ii) a
so-called 'direct' approach [D-I], [F.2]. The variational argument in
[L-T.7] starts from the control problem as the primary issue and
constructs an explicit candidate for the Riccati operator (in terms of
the data of the problem with the help of the optimal solution, see
Remark 5.0), which is then shown to satisfy the ARE (5.1). In
contrast, the direct approach as in [D-I], [F.2] takes the direct study
of well-posedness (existence and uniqueness) of the ARE as the primary
object and only subsequently recovers the control problem (via dynamic
programming) which generates the original ARE. In carrying out its
task, the direct method begins actually with a direct study of the
corresponding Differential (or Integral) Riccati Equation of the
optimal problem over a finite interval [0,T], $T < \infty$, and operates a
limit process as $T \to \infty$ <u>on the</u> Differential Riccati Equation (in line
with a classical approach, which now, however, has to overcome new
technical difficulties, particularly the strong convergence of $B^* P_T(0)$
to $B^* P$). In both approaches, a key point consists in establishing that
the gain operator $B^* P$ (*a priori* not necessarily well defined) is, in
fact, a bounded operator; see (5.3). In the variational approach, this
latter property is accomplished by using analyticity of the free
dynamics, together with a certain 'bootstrap' argument based on the
Young inequality to show that the optimal pair is more regular, indeed
$e^{-\omega t} u^0(t;y_0) \in C([0,\infty];U)$ and $e^{-\omega t} y^0(t;y_0) \in C([0,\infty];Y)$ for $y_0 \in Y$.
(*A priori*, we only know that $u^0 \in L_2(0,\infty;U)$, while a general control
$u \in L_2(0,T;U)$ need <u>not</u> produce in general a corresponding solution
$y \in C([0,T];Y)$; a counterexample being obtained by a parabolic equation
on Ω, with Dirichlet-boundary control where $U = L_2(\Gamma)$, and $Y = L_2(\Omega)$
[Lio.3; p. 217].) All this leads to the regularity property (5.2) via
the explicit representation of P in terms of the optimal solution (see
Remark 5.0), which in turn leads to property (5.3). Instead, in case
of the direct approach [D-I], [F.2], the boundedness (5.3) of the gain
operator $B^* P$ is established by proving first that the solution of the
corresponding Differential Riccati Equation for the problem on [0,T]
possesses the desired regularity properties, and then by passing to the
limit as $T \to \infty$. This, in turn, is accomplished in [F.2] by repeated
applications of the Young's inequality to prove that the optimal

trajectory is in $C([0,T];Y)$ for any $T > 0$; or in [D-I] by a direct study of the evolution equation via a fixed point argument.

Remark 5.1. As remarked in Section 2, Remark 2.4, for the case $T < \infty$, it should be noted that the proof of Theorem 5.1 greatly simplifies and becomes rather straightforward, in fact, in case the constant γ appearing in assumption (1.3) can be taken to be $\gamma < \frac{1}{2}$, or even $\gamma = \frac{1}{2}$, if the operator A is self-adjoint or normal, or has a Riesz basis of eigenvectors. Indeed, in this case, standard analytic estimates give at the outset that <u>any</u> solution y to an $L_2(0,\infty;U)$-control function satisfies in fact the regularity property $y \in C([0,T];Y)$, $\forall\, T > 0$; $e^{-\omega t}y(t;y_0) \in C([0,\infty];Y)$ for $y_0 \in Y$. Thus, such property holds automatically true for the optimal y^0 in this case (while it is a distinctive property of y^0 to be proved when $\frac{1}{2} < \gamma < 1$, not shared by general solutions y to $L_2(0,\infty;U)$-controls u, as discussed above). As a consequence, one obtains immediately then that B^*P is bounded, by using the explicit representation of P in terms of the optimal solution. Our sections 6.1, 6.2, and 6.3 below will concentrate on distinctive, physically relevant, analytic problems where in fact $\frac{1}{2} < \gamma < 1$. Here, Theorem 5.1 will apply, while other treatments such as the one in [PS] cannot cover these cases. ∎

Remark 5.2. The 'detectability' assumption (D.C.) = (5.10) guarantees not only uniqueness of the Riccati solution P to the ARE, but also the property that the resulting feedback semigroup $e^{A_P t}$ is exponentially stable as in (5.11); and, indeed, it is the latter property that is used to prove the former. The exponential decay of $e^{A_P t}$ is a particularly attractive feature in applications, for then the Riccati operator provides, constructively, a stabilizing feedback operator of the free dynamics $\dot{y} = Ay$ which may be, possibly, unstable to begin with. ∎

5.2. <u>Algebraic Riccati Equation for the second class subject to the 'trace' regularity assumption (H.2) = (1.6)</u>

The study of the ARE is more complicated for the class of dynamics subject to assumption (H.2) = (1.6), rather than to assumption (H.1) = (1.5). Indeed, in this case, there is no smoothing effect of the free dynamics which will make up for the unboundedness of the operator B. And, in fact, in most of the interesting situations, the gain operator B^*P is intrinsically unbounded, see Corollaries 5.4, 5.5 below. This feature is in sharp contrast with the 'analytic' situation described in section 5.1, when assumption (H.1) instead is in force. Thus, boundedness of B^*P for the class (H.1) and unboundedness of B^*P for the class (H.2) in the most interesting situations is a distinguishing feature that tells the two cases apart. On the other hand, it should be noted that, in contrast with the situation of section 5.1, hypothesis (H.2) = (1.6) implies by duality the desired regularity $u \in L_2(0,T;U) \to y \in C([0,T];Y)$ (see (1.12)), which under the hypothesis (H.1) = (1.5) case $\frac{1}{2} < \gamma < 1$, is generally false, but can be proved to be true, however, for the optimal pair (u^0, y^0), as remarked in section 5.1. A rather complete theory for the Algebraic Riccati Equation under present assumptions was first given in [L-T.6] in the canonical case of the wave equation, or more generally second-order hyperbolic equations, with Dirichlet control in $L_2(0,T;L_2(\Gamma))$, which was treated, however, by abstract operator-theoretic methods. This treatment was later put fully on an abstract space framework, and complemented by further results, in [FLT].

Theorem 5.2 ([L-T.6], [L-T.9], [FLT]).
I.   **Existence**. For the second class covered by hypothesis (H.2) = (1.6) and subject to the Finite Cost Condition (1.9), there exists a self-adjoint, non-negative solution $0 \leq P = P^* \in \mathcal{L}(Y)$ of the ARE (5.1) such that:

(i) $P \in \mathcal{L}(\mathcal{D}(A),\mathcal{D}(A_P^*)) \cap \mathcal{L}(\mathcal{D}(A_P),\mathcal{D}(A^*))$,   (5.12)

where the operator

$$A_P = A - BB^*P \qquad (5.13)$$

generates a s.c. semigroup on Y; thus, the ARE (5.1) holds true also for all $x,y \in \mathcal{D}(A_P)$;

(ii) $B^*P \in \mathcal{L}(\mathcal{D}(A),U) \cap \mathcal{L}(\mathcal{D}(A_P);U)$;   (5.14)

(iii) $J(u^0,y^0) = (Py_0,y_0)_Y$;   (5.15)

(iv) $u^0(t) = -B^*Py^0(t)$;   (5.16)

where we write $y^0(t) = y^0(t;y_0)$, $u^0(t) = u^0(t;y_0)$, and (5.16) is understood a.e. in t if $y_0 \in Y$; while instead, if $y_0 \in \mathcal{D}(A_P)$,

then (5.14) implies $y^0(t;y_0) \in C([0,T];\mathcal{D}(A_P))$, and by (5.16),
$u^0(t;y_0) \in C([0,T];U)$ for any $T > 0$.

II.   **Uniqueness**. In addition to the assumption of part I, we assume
that the following 'detectability' condition (D.C.) holds true:

(D.C): There exists $K: Z \supset \mathcal{D}(K) \to Y$ densely defined such that

$$\|K^*x\|_Z \leq C[\,|B^*x|_U + \|x\|_Y\,], \; \forall \; x \in \mathcal{D}(B^*) \subset Y, \qquad (5.17)$$

so that the operator

$$A_K = A + KR \quad \text{(interpreted as closed)} \qquad (5.18)$$

is the generator of a s.c. semigroup $e^{A_K t}$ on Y, which is then assumed
to be exponentially stable on Y:

$$\|e^{A_K t}\|_{\mathcal{L}(Y)} \leq M_K e^{-\omega_K t}, \quad t > 0, \qquad (5.19)$$

for some M_K, $\omega_K > 0$. (For $R > 0$, we choose $K = -c^2 R^{-1}$ with constant c
sufficiently large, and the detectability assumption (5.17)–(5.19) is
automatically satisfied.) Then
(a)   the solution P to the ARE (5.1) is unique within the class of
       non-negative self-adjoint operators in $\mathcal{L}(Y)$ which satisfy the
       regularity properties (5.14);
(b)   the s.c. semigroup $e^{A_P t}$ generated by A_P in (5.13) is

       exponentially (uniformly) stable on Y.   ∎

       The proof of Theorem 5.2 is given in [FLT] and follows the
abstract treatment of the canonical case of second order hyperbolic
equations with Dirichlet control [L–T.6]. It is based on a variational
approach. The following comments, which constrast the technical
methodology available in the case of Theorem 5.2 with that available in
the case of Theorem 5.1, apply. A main difference between the two
cases is that, as pointed out in section 3, at present no Differential
Riccati Equation on [0,T] is available under the assumption
(H.2) = (1.6) with no smoothing of the operator R; i.e., with R subject
only to assumption (5.0) of boundedness, in contrast with the situation
available under assumption (H.1) = (1.5) in section 2. Thus, an

approach to the issue of existence of a solution of the ARE which is based on the classical idea of a limiting process as $T \rightarrow \infty$ on the Differential Riccati Equation is out of question (unlike the analytic case of assumption (H.1) = (1.5), in the direct approach [D-I], [F.2], as described in section 5.1). Therefore, a different strategy is now applied [L-T.6], [FLT]. As in the analytic case treatment of [L-T.7] under assumption (H.1), the existence of a solution to the ARE is now obtained under assumption (H.2) through the following steps: (i) First, one constructs an explicit candidate of a solution, in terms of the data of the problem (the optimal solution);   (ii) next, one establishes the necessary regularity properties of such a candidate, as described in (5.12), (5.14), using its explicit representation;   (iii) finally, one verifies that such candidate operator does satisfy the ARE (5.1).

As mentioned in the introduction of section 5.2, it is important to notice that, in contrast with the situation described by Theorem 5.1 under the analyticity assumption (H.1) = (1.5), the gain operator B^*P is now generally unbounded (in the most interesting situations). Indeed, this property follows from the next result, which is the counterpart of Theorem 3.1(viii).

Theorem 5.3. Let the hypotheses (H.2) = (1.6) for the dynamics and the Finite Cost Condition (1.9) hold true, as in Theorem 5.2, part I. In addition, assume the following exact controllability condition:

(E.C.) $\begin{cases} \text{the equation } \dot{y} = A^*y + R^*v \text{ is exactly controllable} \\ \text{in Y from the origin over some } [0,T], \ T < \infty, \\ \text{within the class of } L_2(0,T;Z)\text{-controls } v. \end{cases}$ (5.20)

(We shall say, in short, that the pair $\{A^*, R^*\}$ is exactly controllable.) Then, the solution operator P to the ARE (5.1) guaranteed by Theorem 5.2, part I, is an isomorphism on Y.   ∎

Corollary 5.4. Under the assumptions (H.2) = (1.6), (1.9), and (5.20) of Theorem 5.3, we have: The operator B: $U \supset \mathcal{D}(B) \rightarrow Y$ is bounded if and only if the operator B^*P: from its domain in $Y \rightarrow U$ is bounded.   ∎

From Corollary 5.4, we see that for the second class subject to assumption (H.2) = (1.6), where moreover B is an unbounded operator (the interesting case), the requirement that the gain operator B^*P be bounded runs into conflict with the assumption (5.20) of exact controllability of the pair $\{A^*,R^*\}$. On the other hand, if (i) the original free dynamics e^{At} is a s.c. <u>group</u> uniformly bound for negative times (the case of all the interesting conservative wave and plate and Schrödinger problems, which yield in fact unitary groups) and (ii) the desirable detectability condition D.C. = (5.17)-(5.19) of the pair $\{A,R\}$ holds true, then the pair $\{A^*,R^*\}$ is uniformly (exponentially) stabilizable (by the operator K^* in the notation of (5.17)). Then, a well-known result of D. Russell (1973) [Ru.2] for time-reversible systems implies that the pair $\{A^*,R^*\}$ is exactly controllable (to the origin, or equivalently, from the origin). Thus:

Corollary 5.5. Assume hypothesis (H.2) = (1.6) for the dynamics, as well as the Finite Cost Condition (1.9) and the Detectability Condition (5.17)-(5.19). Assume, further, that the free dynamics e^{At} is an s.c. group uniformly bounded for negative times. Then the conclusion of Corollary 5.4 applies: B is bounded if and only if B^*P is bounded. ∎

Remark 5.3. The following reference [P-S] also deals with the Algebraic Riccati Equation under the abstract hypothesis (H.2) = (1.6). In addition, however, [P-S] makes the following two further assumptions:

(i) an assumption of the smoothness on the observation operator $R \in \mathcal{L}(Y,Z)$ as expressed by the requirement

$$\int_0^T \|Re^{At}x\|_Z^2 dt \leq c_T\|x\|_V^2 \tag{5.21}$$

where V is a space strictly larger than Y and with weaker topology than Y, such that $B \in \mathcal{L}(U,V)$ and e^{At} generates a s.c. semigroup on both Y and V.

(ii) The assumption that the Finite Cost Condition holds true for all initial data in the strictly larger space V, not only on Y. (This assumption is <u>not</u> true in the cases of conservative waves and plates problems such as those of sections 6.2, 6.3, and 6.4). Under the above assumptions, [P-S] claims existence and, under the additional Detectability Condition on V, uniqueness of the solution P of the ARE, where P enjoys the following regularity property:

$P \in \mathcal{L}(V,V')$, V' = dual of V with respect to Y-topology, $\qquad$ (5.22) which in turn implies boundedness of the gain operator

$$B^*P \in \mathcal{L}(Y,U). \qquad (5.23)$$

In view of the above Corollaries 5.4, 5.5, we can say that: not only are the results of [P-S] on the ARE subsumed by the earlier Theorem 5.2, first given in the canonical case of the wave equation by abstract operator methods [L-T.6], and then fully cast in abstract setting in [FLT]; moreover, and in contrast with Theorem 5.2, the results of [P-S] cannot cover the important class of conservative waves and plates problems with B unbounded (point or boundary control) and R > 0, which is precisely the class which offers a main justification for introducing the abstract assumption (H.2) = (1.6) in the first place.

In addition, the results of [PS] cannot cover the distinctive, physically relevant, analytic (parabolic) problems such as those of our section 6.1, 6.2, and 6.3 below, as the regularity required by [PS] is violated ($\frac{1}{2} < \gamma < 1$: see Remark 5.1).

If one is interested in situations where the gain operator B^*P is bounded as in (5.23), then an assumption on the smoothing of the observation R much weaker than (5.21) in [P-S] suffices to achieve this goal. Indeed, the following result holds true.

Theorem 5.6. In addition to the hypotheses of Theorem 5.2, assume the following regularity property on R:

$$\int_0^\infty \|R^*Re^{-\hat{A}t}Bu\|_Y dt \leq C\|u\|_U, \quad u \in U \qquad (5.21bis)$$

where $-\hat{A}$ is the translation of A introduced above (5.0), which generates an s.c. semigroup $e^{-\hat{A}t}$ uniformly stable: $\|e^{-\hat{A}t}\| \leq \hat{M}e^{-\hat{\omega}t}$, $\hat{\omega} > 0$, $t \geq 0$.

Then the operator P, in addition to the properties guaranteed by Theorem 5.2, satisfies (5.23): $B^*P \in \mathcal{L}(Y,U)$. ∎

Hypothesis (5.21bis) guarantees that the corresponding $P_T(0)$ satisfies the Differential Riccati Equation for all T > 0, see Section 3.3. Then, a limiting argument on the formula defining P yields Theorem 5.6.

Since in the treatment of [P-S], we have $B \in \mathcal{L}(U,V)$ by assumption, then condition (5.21) in [P-S] implies a *fortiori* that

$$\int_0^T \|Re^{At}Bu\|_Z^2 dt \le C_T \|u\|_U^2, \quad u \in U \qquad (5.21tris)$$

A comparison between our condition (5.21bis) and condition (5.21tris) reveals that (5.21bis) is weaker than (5.21tris) on a few grounds:

(i)   (5.21bis) requires only an L_1 condition on any finite T, while (5.21tris) requires an L_2 condition in time;

(ii)   condition (5.21bis) uses the smoothing of R^*R ("twice" that of the original R), while condition (5.21tris) needs the smoothing only of R; in other words, even within an L_2-test, condition (5.21tris) requires "twice as much" smoothing of R than condition (5.21bis) does.

The next result is the counterpart of the Dual Differential Riccati Equation (3.37) for $T < \infty$.

<u>Corollary 5.7</u>. [F-L-T] Under the assumptions of Theorem 5.3 which guarantee that the Riccati operator P is an isomorphism on Y, set $Q = P^{-1} \in \mathcal{L}(Y)$. Then, Q satisfies the following Dual Algebraic Riccati Equation (DARE):

$$(AQx, y)_Y + (QA^*x, y)_Y + (RQx, RQy)_Z - (B^*x, B^*y)_U = 0,$$
$$\forall x, y \in \mathcal{D}(A^*) \subset \mathcal{D}(B^*) \subset Y; \qquad (5.24)$$

$$Q \in \mathcal{L}(\mathcal{D}(A^*); \mathcal{D}(A_p)) \cap \mathcal{L}(\mathcal{D}(A_p^*); \mathcal{D}(A)), \quad A_p = A - BB^*P. \quad \blacksquare \qquad (5.25)$$

Equation (5.24) will be henceforth referred to as Dual Algebraic Riccati Equation with respect to the (original) Algebraic Riccati Equation (5.1). A comparison between (5.24) and (5.1) reveals the following correspondence (Table 5.1).

TABLE 5.1. Correspondence Between Original and Dual ARE

	A	A^*	R	B	B^*	P	Z
Original ARE (5.1)	A	A^*	R	B	B^*	P	Z
Dual ARE (5.24)	$-A^*$	$-A$	B^*	R^*	R	Q	U

Thus, to the original dynamics (1.1) and to its corresponding (infinite horizon) control problem (1.2), there corresponds the dual dynamics and its corresponding control problem indicate below.

TABLE 5.2. Original and Dual Problem

Original Problem	Dual Problem
dynamics (1.1):	dynamics:
$\dot{y}$ = Ay+Bu on Y;	$\dot{z}$ = $-A^*z+R^*v$ on Y;
Cost (1.2):	Cost:
$J(u,y) = \int_0^\infty \{\|Ry(t)\|_Z^2 + \|u(t)\|_U^2\}dt$	$J(v,z) = \int_0^\infty \{\|B^*z(t)\|_U^2 + \|v(t)\|_Z^2\}dt$

From the correspondence of Table 5.2, we see plainly that the DARE (5.24) is associated to the dynamics $\dot{z}$ = $-A^*z+R^*v$, whose well-posedness, however, requires *the additional assumption that* $-A^*$ *(equivalently,* $-A$*) be the generator of a s.c. semigroup on* Y; *i.e., that* A^* *(equivalently,* A*) be the generator of an s.c. group on* Y. As a consequence of this assumption and of hypothesis (H.2) = (1.6), it may be shown [F-L-T, Lemma 7.0] that B^*z is a well-defined element of $L_2(0,T;U)$ for any T > 0, when $v \in L_2(0,T;Z)$. In the following results we shall discuss the DARE under the assumption that A be an s.c. group generator, a special but important case (this hypothesis is automatically fulfilled in the case of, say, conservative waves and plates equations).

It was already noted in section 3 as well as in the paragraph below Theorem 5.2 that no Differential Riccati Equation on [0,T] is at present available under the assumption (H.2) = (1.6) with no smoothing of the operator R, i.e., with R subject only to assumption (5.0) of boundedness. In such generality, it is still true that the Riccati operator P of Theorem 5.2 can be identified by the limit relation,

$$Px = \lim_{T\uparrow\infty} P_T(0)x, \quad x \in Y, \tag{5.26}$$

where $P_T(t) \in \mathcal{L}(Y)$, $0 \le t \le T$, is the operator in (3.6) which is explicitly defined in terms of the system's data via the corresponding optimal solution of the quadratic problem over [0,T]; moreover, such $P_T(t)$ does realize the pointwise (a.e.) feedback synthesis,

$$u_T^0(t,0;y_0) = -B^* P_T(t) y_T^0(t,0;y_0), \quad \text{a.e. in } [0,T], \quad (5.27)$$

of the optimal pair $\{u_T^0, y_T^0\}$ as seen in Theorem 3.1 as well as some further relations typical of the Riccati theory, noted there and in Section 3.4. However, what is missing in such generality for $R \in \mathcal{L}(Y,Z)$ is a claim that $P_T(t)$ satisfies the Differential Riccati Equation. In what follows, we shall show that when the dynamics is time reversible (A generates an s.c. group), it is possible to make a connection, under natural assumptions, between the algebraic Riccati operator P and the solution to 'some' Differential Riccati Equation, indeed, the Dual Differential Riccati Equation introduced below in (5.30).

Theorem 5.8. ([FLT], [F.3]) Let A be an s.c. group generator on Y and assume hypothesis (H.2) = (1.6) for the dynamics. Finally, assume the Finite Cost Condition for the Dual Problem in Table 5.2:

$$\left\{ \begin{array}{l} \text{For each } z_0 \in Y, \text{ there exists } v \in L_2(0,\infty;Z) \text{ such} \\ \text{that } J(v,z) < \infty, \text{ where } z \text{ is the solution due to } v. \end{array} \right\} \quad (5.28)$$

Then:

(i) there exists an operator $\hat{Q} \in \mathcal{L}(Y)$, $\hat{Q} = \hat{Q}^* \geq 0$, which satisfies the DARE (5.24);

(ii) if, in addition, the pair $\{-A,B\}$ (equivalently, the pair $\{A,B\}$ is exactly controllable over some interval [0,T] within the class of $L_2(0,T;U)$-controls, then the DARE (5.24) admits a unique solution, given by $\hat{Q}$, in the class of all $Q \in \mathcal{L}(Y)$, such that $Q = Q^* \geq 0$;

(iii) $\hat{Q}$ is given by the strong limit

$$\hat{Q}x = \lim_{T\uparrow\infty} Q_T(0)x, \quad x \in Y, \quad (5.29)$$

where $Q_T(t) \in \mathcal{L}(Y)$, $0 \leq t \leq T$, is the unique solution of the following Dual Differential Riccati Equation

$$
\begin{cases}
\dfrac{d}{dt}\,(Q_T(t)x,y)_Y = (Q_T(t)x,A^*y)_Y + (A^*x,Q_T(t)y)_Y \\
\qquad\qquad + (RQ_T(t)x,RQ_T(t)y)_Z - (B^*x,B^*y)_U \\
\qquad\qquad \text{for all } x,y \in \mathcal{D}(A^*), \\
Q_T(T) = 0,
\end{cases}
\qquad (5.30)
$$

uniqueness being in the class of operators $Q(\cdot) \in \mathcal{L}(Y;C([0,T];Y)$ such that $(Q(t)x,y)$ is differentiable in t for each $x,y \in \mathcal{D}(A^*)$;

(iv) the pair $\{-A,B\}$ (equivalently, the pair $\{A,B\}$ is exactly controllable on some $[0,T]$ within the class of $L_2(0,T;U)$-controls, if and only if $Q_T(0)$ is an isomorphism on Y, in which case $\hat{Q}$ is an isomorphism on Y as well. ∎

The proof of Theorem 5.8 is in [FLT, Theorems 2.6 and 2.7], where further results may be found. It should be noted that the analysis of the DARE (5.24), as well as its derivation starting from the Dual control problem in Table 5.2, are a much simpler task than the analysis of the original ARE (5.1) and its derivation starting from the original control problem (1.1), (1.2). Indeed, the quadratic term $(RQx,RQy)_Z$ in (5.24) in the unknown Q occurs with the bounded operator R, while the quadratic term $(B^*Px,B^*Py)_U$ in (5.1) in the unknown P occurs with the unbounded operator B^*. The well-posedness (existence and uniqueness) of the DARE for Q can be readily handled by arguments by now standard (fixed point plus a priori bounds) [F.3], [FLT], following the original treatment in [DaP].

The same considerations applied to the original ARE (5.1) say that the case of R unbounded and B bounded is much easier than the case of R bounded and B unbounded; compare the term $(Rx,Ry)_Y$ with the term $(B^*Px,B^*Py)_U$ in (5.1).

Theorem 5.8 states that the operator $\hat{Q}$ defined by the limit (5.29) of solutions to the Dual Differential Riccati Equation (5.30) satisfies the same Dual Algebraic Riccati Equation (5.24) that the operator P^{-1} --if it exists!--would satisfy, see Corollary 5.7. Since under exact controllability of the pair $\{-A,B\}$, the operator $\hat{Q}$ is an isomorphism on Y, the question arises as to whether or when the analysis of the original control problem leading to the Riccati operator P, and the analysis of the dual control problem leading to the dual Riccati operator $\hat{Q}$ merge, i.e., more precisely as to whether or when we have $\hat{Q} = P^{-1}$, or $\hat{Q}^{-1} = P$. In general, the answer is in the negative, see Example below. Indeed, the very identification of P with $\hat{Q}^{-1}$ requires that P be an isomorphism on Y. It is most gratifying therefore that the identification $P = \hat{Q}^{-1}$ holds true (when A is a group generator) provided that both pairs $\{-A, B\}$ and $\{A^*, R^*\}$ are exactly controllable on some $[0, T]$, $T < \infty$, i.e., precisely the conditions under which $\hat{Q}$ and P are both isomorphisms, see Theorem 5.3 and Theorem 5.7(iv), respectively: this is the content of Theorem 5.9 below.

Example. [FLT, p.325] Let $R = 0$, $B \in \mathcal{L}(U, Y)$, $-A^*$ stable, and $\{A, B\}$ exactly controllable over some $[0, T]$. Then, trivially, $P = 0$. On the other hand, the Finite Cost Condition (5.28) for the dual problem is satisfied since $-A^*$ is stable and

$$(\hat{Q}x, y)_Y = \int_0^\infty (B^* e^{-A^* t} x, B^* e^{-A^* t} y)_U dt, \qquad (5.31)$$

where $\hat{Q}$ satisfies the DARE. Moreover, $\hat{Q}$ is an isomorphism on Y, by Theorem 5.8(iv): $\hat{Q}^{-1} \in \mathcal{L}(Y)$. However, $\hat{Q}^{-1} \neq P$. ∎

Theorem 5.9. [FLT, Theorem 2.7] Let A generate an s.c. group on Y. If both pairs $\{A, B\}$ and $\{A^*, R^*\}$ are exactly controllable over some $[0, T]$, then

$$P^{-1} = \hat{Q}. \qquad \blacksquare \qquad (5.32)$$

Combining Theorems 5.8 and 5.9, we obtain

Corollary 5.10. Under assumption (H.2), let A generate an s.c. group on Y, and let $\{A, B\}$ and $\{A^*, R^*\}$ be both exactly controllable on some $[0, T]$. Then

$$Px = \lim_{T \uparrow \infty} Q_T^{-1}(0)x, \qquad x \in Y, \qquad (5.33)$$

where $Q_T(t)$, $0 \leq t \leq T$, is the unique solution to the Dual Differential Riccati Equation (5.30). ∎

The above Corollary 5.10 characterizes the Riccati operator R as a strong limit of solutions to the DDRE (5.30), as desired.

The accompanying diagram illustrates a few main points of the original and dual problem, and their merging at the level of establishing Q defined by $Q = P^{-1}$ coincides with $\hat{Q}$ defined by (5.29).

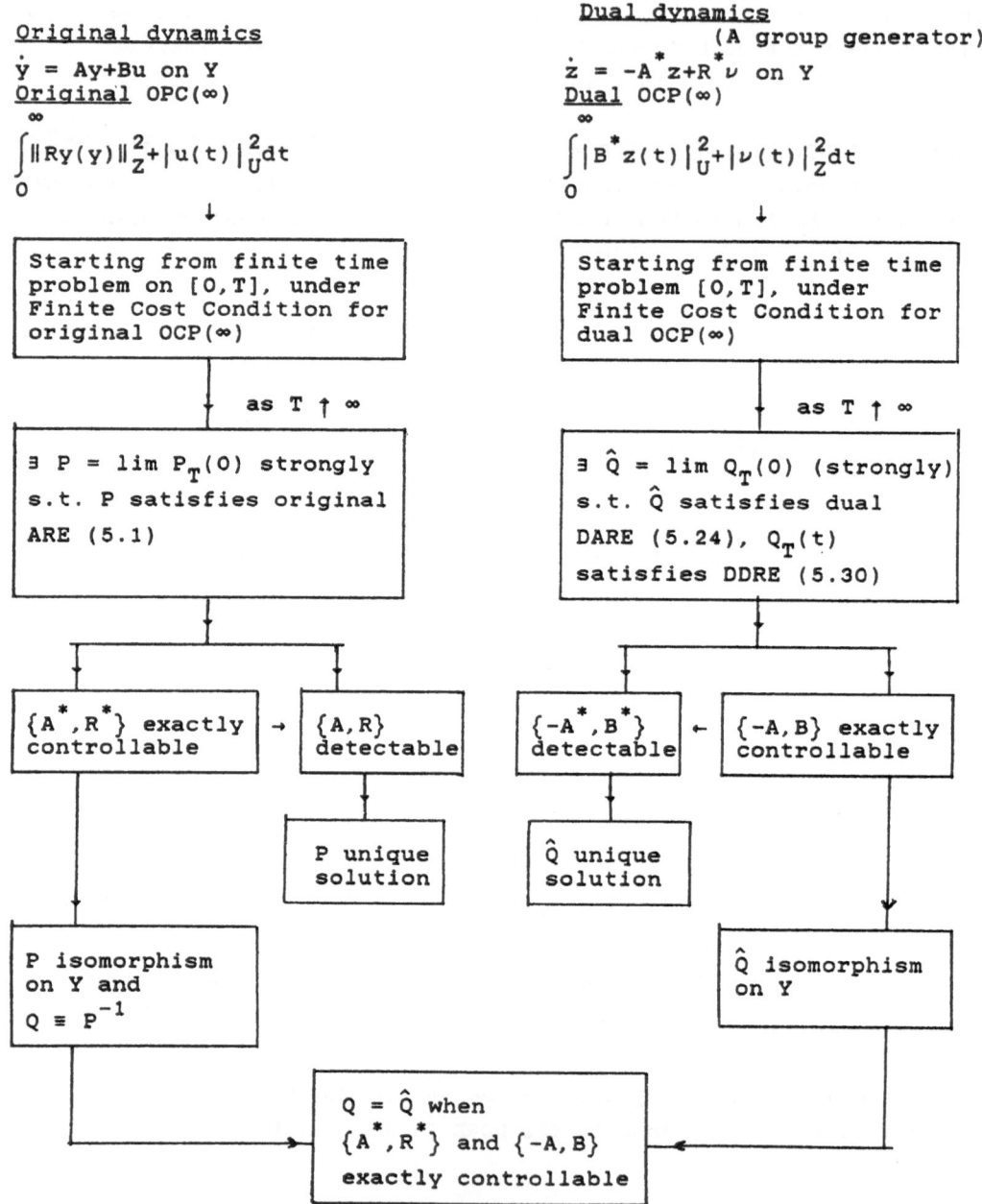

6.   **Examples of partial differential equation problems satisfying**
        (H.1)

        In this section, we illustrate the applicability of Theorems 2.1
($T < \infty$) and 5.1 ($T = \infty$) for the 'analytic' class subject to hypothesis
(H.1) = (1.5). In passing, some p.d.e. problems will be exhibited
which satisfy both (H.1) and (H.2). Obvious candidates for the
analytic class are heat or diffusion problems. A few canonical cases
thereof will be treated in sections 6.1 and 6.2 below, following e.g.,
[Las.4], [L-T.7]. In section 6.3, we shall then analyze examples of
plates with a strong degree of damping (structural damping), such as
may arise in the study of flexible structures. All examples in
sections 6.1, 6.2, 7.1, 7.2, 7.3, and most of those in section 6.3, are
not covered by other treatments such as the one in [PS].

6.1.   **Class (H.1): Heat equation with Dirichlet boundary control**
        Let $\Omega \subset R^n$ be an open bounded domain with sufficiently smooth
boundary Γ. In Ω, we consider the Dirichlet mixed problem for the heat
equation in the unknown $y(t,x)$:

$$\begin{cases} y_t = \Delta y + c^2 y & \text{in } (0,T] \times \Omega \equiv Q, & (6.1a) \\ y(0,\cdot) = y_0 & \text{in } \Omega, & (6.1b) \\ y|_\Sigma = u & \text{in } (0,T] \times \Gamma \equiv \Sigma, & (6.1c) \end{cases}$$

with boundary control $u \in L_2(\Sigma)$ and $y_0 \in L_2(\Omega)$. We explicitly consider
$T = \infty$. The cost functional which we wish to minimize is then

$$J(u,y) = \int_0^\infty \{\|y(t)\|^2_{L_2(\Omega)} + \|u(t)\|^2_{L_2(\Gamma)}\} \, dt. \qquad (6.2)$$

        Note that the above problem (6.1), (6.2) does not belong to the
class (H.2). In fact, with, say, $y_0 = 0$ and $u \in L_2(\Sigma)$, we only have
$y \in H^{\frac{1}{2},\frac{1}{4}}(Q)$, but y does not belong to $C([0,T];L_2(\Omega))$, even in
1-dimension.

Abstract setting [B.1], [W], [Las.4], [T.6]. To put problem (6.1),
(6.2) into the abstract setting of the preceding sections, we introduce
the operator

$$Ah = \Delta h + c^2 h; \quad \mathcal{D}(A) = H^2(\Omega) \cap H^1_0(\Omega), \qquad (6.3)$$

select the spaces

$$Z = Y = L_2(\Omega); \qquad U = L_2(\Gamma), \qquad (6.4)$$

and finally define the operators

$$Bu = -AD_1 u; \qquad R = I, \qquad (6.5)$$

where D_1 (Dirichlet map) is defined by

$$h = D_1 g \quad iff \quad (\Delta + c^2)h = 0 \text{ in } \Omega; \qquad h_{|\Gamma} = g, \qquad (6.6)$$

and by elliptic theory and [Gr],

$$D_1: \text{ continuous } L_2(\Gamma) \to H^{\frac{1}{2}}(\Omega) \subset H^{\frac{1}{2} - 2\varepsilon}(\Omega) \equiv \mathcal{D}(A_D^{\frac{1}{4} - \varepsilon}), \quad \forall \ \varepsilon > 0, \qquad (6.7)$$

$$A_D h = -\Delta h, \qquad \mathcal{D}(A_D) = H^2(\Omega) \cap H_0^1(\Omega), \qquad (6.8)$$

In (6.5) A is the isomorphic extension of A in (6.3), from, say, $L_2(\Omega) \to [\mathcal{D}(A)]'$.

Assumption (1.3): $(\hat{A})^{-\gamma} B \in \mathcal{L}(U, Y)$. Assumption (1.3) is satisfied in our present case with $\gamma = \frac{1}{4} + \varepsilon$, $\forall \ \varepsilon > 0$. In fact, we may take $\hat{A} = A_D$. From (6.7), we have

$$AD: \text{ continuous } L_2(\Gamma) \to [\mathcal{D}(\hat{A}^{\frac{1}{4} + \varepsilon})]' = [\mathcal{D}(A_D^{\frac{1}{4} + \varepsilon})]', \qquad (6.9)$$

and we then have with $\gamma = \frac{1}{4} + \varepsilon$ via (6.5), that our claim is verified,

$$\hat{A}^{-\gamma} B = -A_D^{-\gamma} AD \in \mathcal{L}(L_2(\Gamma); L_2(\Omega)) = \mathcal{L}(U; Y). \qquad (6.10)$$

Assumption (H.1) = (1.5). The operator A in (6.3) generates an s.c. semigroup e^{At}, on $L_2(\Omega)$, which is moreover analytic here for $t > 0$ (and contraction after a suitable translation of the generator).

Finite Cost Condition (1.9). The generator A has (for suitably large constant c^2 in (6.3)) only finitely many eigenvalues of finite multiplicity, since its resolvent is compact and e^{At} is analytic. Thus, the stabilization theory as in [T.1], [T.5], [L-T.17], [M-T, Appendix], etc. applies: The problem is stabilizable on $L_2(\Omega)$ if and only if its projection onto the finite dimensional unstable subspace is controllable. In particular, as shown in [T.5], [L-T.17], one may prescribe the stabilizing feedback to be of the form

$$u(t) = \sum_{n=1}^{N} (y(t), w_n)_{L_2(\Omega)} g_n \qquad (6.11)$$

for suitable vectors $w_k \in L_2(\Omega)$ and $g_k \in L_2(\Gamma)$ and suitable (minimal) N as described there, in order to stabilize uniformly the corresponding feedback system in the norm of $H^{\frac{1}{2}-\varepsilon}(\Omega)$ in fact. Thus, a fortiori the Finite Cost Condition on $L_2(\Omega)$ is satisfied.

__Detectability Condition (5.10)__. This is automatically satisfied since in our case R = I, see (6.5).

__Conclusion__: T = ∞. Theorem 5.1 applies to problem (6.1), (6.2) [L-T.7].

__Conclusion__: T < ∞. Theorem 2.1 applies to problem (6.1) and (6.2) with T < ∞ for any final state operator G that makes GL a closed (closable) operator. See Remark 2.1, in particular the (only) sufficient condition (2.25).

__Remark 6.0__. The above analysis applies, with no essential change, to y(t) penalized in $L_2(0,T;H^{\frac{1}{2}-\varepsilon}))$, with $y_0 \in H^{\frac{1}{2}-\varepsilon}(\Omega)$, rather than in $L_2(0,T;L_2(\Omega)$ as in (6.2) where then $\gamma = 1-\varepsilon/2$. ∎

6.2. __Class (H.1): Heat equation with Neuman boundary control__

Now we consider problem (6.1a-b) with (6.1c) replaced now by

$$\frac{\partial y}{\partial \nu}\Big|_\Sigma = u, \qquad \text{on } \Sigma, \qquad (6.12)$$

$u \in L_2(\Sigma)$ and cost functional for T = ∞:

$$J(u,y) = \int_0^\infty \{\|y(t)\|^2_{L_2(\Omega)} + \||\nabla y(t)|\|^2_{L_2(\Omega)} + \|u(t)\|^2_{L_2(\Gamma)}\} \, dt, \qquad (6.13)$$

instead of (6.2), where we take $y_0 \in H^1(\Omega)$. We note again that the present problem (6.1a-b), (6.12), (6.13), does not belong to the class (H.2). This is so since, say, with $y_0 = 0$ and $u \in L_2(\Sigma)$, we have

$$y \in H^{\frac{3}{4},\frac{3}{4}}(Q), \text{ hence } y \in C([0,T];H^{\frac{1}{2}}(\Omega)), \tag{6.14}$$

but y does not belong to $C([0,T];H^1(\Omega))$.

Abstract setting. To put problem (6.1a-b), (6.12), (6.13) into the abstract setting, we introduce the operator

$$Ah = \Delta h + c^2 h; \qquad \mathcal{D}(A) = \{h \in H^2(\Omega), \left.\frac{\partial h}{\partial \nu}\right|_\Gamma = 0\}, \tag{6.15a}$$

which we shall consider as lifted to

$$A: \mathcal{D}(\hat{A}^{\frac{3}{2}}) \to \mathcal{D}(\hat{A}^{\frac{1}{2}}) = H^1(\Omega), \tag{6.15b}$$

and select the spaces and operators

$$Z = Y = H^1(\Omega) = \mathcal{D}(\hat{A}^{\frac{1}{2}}); \qquad U = L_2(\Gamma), \tag{6.16}$$

$$Bu = -ANu, \qquad R = I, \tag{6.17}$$

with A in (6.17) the isomorphic extension, say, $L_2(\Omega) \to [\mathcal{D}(A)]'$ of A in (6.15a). Here, without loss of generality, we assume that $-c^2$ is not an eigenvalue of Δ with homogeneous Neumann B.C., so that A is boundedly invertible on $L_2(\Omega)$ and the Neumann map N is well defined by

$$h = Ng \quad \text{iff} \quad (\Delta + c^2)h = 0 \text{ in } \Omega; \quad \left.\frac{\partial h}{\partial \nu}\right|_\Gamma = g. \tag{6.18}$$

We have from elliptic theory and [Gr],

$$N: \text{continuous } L_2(\Gamma) \to H^{\frac{3}{2}}(\Omega) \subset H^{\frac{3}{2}-2\varepsilon}(\Omega) = \mathcal{D}(\hat{A}^{\frac{3}{4}-\varepsilon}), \quad \forall \varepsilon > 0, \tag{6.19}$$

Assumption (1.3): $(\hat{A})^{-\gamma}B \in \mathcal{L}(U;Y)$. Assumption (1.3) holds true in the present case with $\gamma = \frac{3}{4}+\varepsilon$, $\forall \varepsilon > 0$. In fact, with $\gamma = \frac{3}{4}+\varepsilon$, we need to show that

$$\hat{A}^{-\gamma}B \in \mathcal{L}(U,Y) = \mathcal{L}(L_2(\Gamma),\mathcal{D}(\hat{A}^{\frac{1}{2}})), \tag{6.20}$$

equivalently that (see (6.17)),

$$\hat{A}^{\frac{1}{2}}A^{-\gamma}AN = \hat{A}^{\frac{1}{2}-\gamma+\frac{3}{4}+\varepsilon}\hat{A}^{\frac{3}{4}-\varepsilon}N \in \mathcal{L}(L_2(\Gamma),L_2(\Omega)), \tag{6.21}$$

which is precisely true in view of (6.19) since $\frac{1}{2}-\gamma+\frac{3}{4}+\varepsilon = 0$.

Assumption (H.1) = (1.5). Since A defined in (6.15a) generates an s.c. analytic semigroup on $L_2(\Omega)$, then its lifting as in (6.15b) generates an s.c. analytic semigroup on $\mathcal{D}(\hat{A}^{\frac{1}{2}}) = H^1(\Omega)$ as desired.

Finite Cost Condition (1.9). Consideration similar to those made for the Dirichlet case apply now; see e.g., [T.5], [L-T.17] for uniform stabilization results in $H^{\frac{1}{2}-\varepsilon}(\Omega)$ in fact. Thus, *a fortiori* the Finite Cost Condition (1.9) on $H^1(\Omega)$ holds true for problem (6.1a-b), (6.12), (6.13).

Detectability Condition (5.10). With R = I, this is automatically satisfied.

Conclusion: T = ∞. Theorem 5.1 applies to problem (6.1a-b), (6.12), (6.13).

Conclusion: T < ∞. Theorem 2.1 applies to problem (6.1a-b), (6.12), (6.13) with any final state operator that makes GL closed (closable). See Remark 2.1, in particular the (only) sufficient condition (2.25).

We also remark that the above analysis applies, with no essential change, to y(t) penalized in $L_2(0,T;H^{\frac{1}{2}-\varepsilon}(\Omega))$ with $y_0 \in H^{\frac{1}{2}-\varepsilon}(\Omega)$ rather than in $L_2(0,T;H^1(\Omega))$ as in (6.13), where then $\gamma = 1-\varepsilon/2$. Here we can take $G = \hat{A}^{\frac{1}{2}-\varepsilon/2}$ with $Z = L_2(\Omega)$. ∎

Remark 6.1. The choice of the functional

$$J(u,y) = \int_0^\infty \|y(t)\|^2_{L_2(\Omega)} + \|u(t)\|^2_{L_2(\Gamma)}\} \, dt, \qquad (6.22)$$

in place of (6.13) considerably simplifies the analysis, since with $Y = L_2(\Omega)$ one easily sees now that in this case we have that assumption (1.3) holds true with $\gamma = \frac{1}{4}+\varepsilon < \frac{1}{2}$. Then, Remark 5.1 applies. This easier problem belongs also to the class (H.2). Thus both Theorems 5.1 and 5.2 are applicable, but Theorem 5.1 is plainly to be preferred. ∎

Remark 6.2. Having solved in $H^1(\Omega)$ the quadratic cost problem for the heat equation problem (6.1a-b), (6.12), (6.13) (indeed, as remarked in $H^{\frac{1}{2}-\varepsilon}(\Omega)$, if we like), we can then obtain as a consequence a solution to the "purely boundary" quadratic cost problem which penalizes the cost functional

$$J(u,y) = \int_0^\infty \|y(t)|_\Gamma\|^2_{L_2(\Gamma)} + \|u(t)\|^2_{L_2(\Gamma)}\} \, dt \qquad (6.23)$$

over all $u \in L_2(0,\infty;L_2(\Gamma))$ with y solution to (6.1a-b), (6.12), and $y_0 \in H^1(\Omega)$. Now we take $Y = H^1(\Omega)$, $Z = L_2(\Gamma)$ and R is the (Dirichlet) trace operator $y \to Ry = y|_\Gamma$: continuous $H^1(\Omega) \to H^{\frac{1}{2}}(\Gamma))$. The previously recalled uniform stabilization results for the solution y in $H^1(\Omega)$ of the corresponding feedback closed loop problem with u, say, of the form as in (6.11), guarantees *a fortiori* exponential uniform decay of $y(t)|_\Gamma$ in $H^{\frac{1}{2}}(\Gamma)$ in fact. Thus, the required Finite Cost Condition (1.9) for (6.23) is satisfied. Moreover, in order to satisfy the Detectability Condition (5.10) we appeal to the stabilization results as in [T.6], see also [L-T.18], to obtain the required "stabilizing" operator $K \in \mathcal{L}(L_2(\Gamma),H^1(\Omega))$ in (5.10), which may be taken of the form

$$K\cdot = \sum_{n=1}^N (\cdot|_\Gamma, w_n)_{L_2(\Gamma)} g_n$$

for suitable $w_n \in L_2(\Gamma)$ and $g_n \in H^1(\Omega)$. If such K is added to the heat equation with homogeneous boundary conditions,

$$\begin{cases} y_t = (\Delta+c^2)y + \sum_{n=1}^N (y|_\Gamma, w_n)_{L_2(\Gamma)} g_n & \text{in Q,} & (6.24c) \\[2mm] y(0,\cdot) = y_0 & \text{in } \Omega, & (6.24b) \\[2mm] \dfrac{\partial y}{\partial \nu}\Big|_\Sigma \equiv 0 & \text{in } \Sigma, & (6.24c) \end{cases}$$

then under suitable conditions on w_n, g_n, uniform decay in (at least) $H^1(\Omega)$ will result [T.6], as desired. The penalization $y(t)|_\Sigma \in L_2(0,T;L_2(\Gamma))$ in (6.33) can be pushed, for the above analysis to go through with no essential change, to $y(t)|_\Sigma \in L_2(0,T;H^{1-\varepsilon}(\Gamma))$ if we take $y_0 \in H^{\frac{1}{2}-\varepsilon}(\Omega)$. ∎

6.3. Class (H.1): Structurally damped plates with point control or boundary control

Example 6.1. **The case** $\alpha = \frac{1}{2}$ in [C-R], [C-T.1-2]. Consider the following model of a plate equation in the deflection $w(t,x)$, where $\rho > 0$ is any constant:

$$
\begin{cases}
w_{tt} + \Delta^2 w - \rho \Delta w_t = \delta(x-x^0)u(t) & \text{in } (0,T] \times \Omega = Q, & (6.25a) \\
w(0, \cdot) = w_0; \quad w_t(0, \cdot) = w_1 & \text{in } \Omega, & (6.25b) \\
w|_\Sigma \equiv \Delta w|_\Sigma \equiv 0 & \text{in } (0,T] \times \Gamma = \Sigma, & (6.25c)
\end{cases}
$$

with load concentrated at the interior point x^0 of an open bounded (smooth) domain Ω of R^n, $n \leq 3$. Regularity results for problem (6.25), and other problems of this type, are given in [T.4]. Consistently with these results, the cost functional we wish to minimize is for $T = \infty$:

$$
J(u,w) = \int_0^\infty \{ \|w(t)\|_{H^2(\Omega)}^2 + \|w_t(t)\|_{L_2(\Omega)}^2 + \|u(t)\|_{L_2(\Omega)}^2 \} \, dt, \quad (6.26)
$$

where $\{w_0, w_1\} \in [H^2(\Omega) \cap H_0^1(\Omega)] \times L_2(\Omega)$.

Abstract setting. To put problems (6.25), (6.26) into the abstract setting of the preceding sections, we introduce the strictly positive definite operator

$$
Ah = \Delta^2 h; \quad \mathcal{D}(A) = \{h \in H^4(\Omega): h|_\Gamma = \Delta h|_\Gamma = 0\} \quad (6.27)
$$

and select the spaces and operators

$$
Y \equiv \mathcal{D}(A^{\frac{1}{2}}) \times L_2(\Omega) = [H^2(\Omega) \cap H_0^1(\Omega)] \times L_2(\Omega); \quad U = R^1, \quad (6.28)
$$

$$
A = \begin{vmatrix} 0 & I \\ -A & -\rho A^{\frac{1}{2}} \end{vmatrix}; \quad Bu = \begin{vmatrix} 0 \\ \delta(x-x^0)u \end{vmatrix}; \quad R = I \quad (6.29)
$$

to obtain the abstract model (1.1), (1.2). We need to verify a few assumptions.

Assumption (1.3): $(-A)^{-\gamma} B \in \mathcal{L}(U,Y)$. It is easy to verify that assumption (1.3) is satisfied with $\gamma = 1$. Indeed, from (6.29), we require that

$$(-A)^{-1}Bu = \begin{vmatrix} \rho A^{-\frac{1}{2}} & A^{-1} \\ -I & 0 \end{vmatrix} \begin{vmatrix} 0 \\ \delta(x-x^0)u \end{vmatrix} = \begin{vmatrix} A^{-1}\delta(x-x^0)u \\ 0 \end{vmatrix} \in Y, \qquad (6.30)$$

i.e., from (6.28), we require that $A^{-\frac{1}{2}}\delta(x-x^0) \in L_2(\Omega)$, or that (#): $\delta(x-x^0) \in [\mathcal{D}(A^{\frac{1}{2}})]'$, the dual of $\mathcal{D}(A^{\frac{1}{2}})$ with respect to $L_2(\Omega)$. Since it is true that $\mathcal{D}(A^{\frac{1}{2}}) \subset H^2(\Omega)$ for the fourth order operator A in (6.27) (in fact, regardless of the particular boundary conditions), and thus $[H^2(\Omega)]' \subset [\mathcal{D}(A^{\frac{1}{2}})]'$, then condition (#) is satisfied provided $\delta(x-x^0) \in [H^2(\Omega)]'$, i.e., provided $H^2(\Omega) \subset C(\bar{\Omega})$, which is indeed the case by Sobolev embedding provided $2 > \frac{n}{2}$, or $n < 4$, as required.

However, the above result is not sufficient for our purposes as--according to assumption (H.1) = (1.5)--we need to show that we can take $\gamma < 1$ in (1.3). As a matter of fact, we now show that assumption (1.3) holds true for any $\gamma > \frac{n}{4}$, which then for $n \leq 3$ yields $\gamma < 1$ as desired. To this end, we note that

$$(-A)^{-\gamma}B \in \mathcal{L}(U,Y) \text{ if and only if } B \in \mathcal{L}(U,[\mathcal{D}((-A^*)^\gamma)]' \qquad (6.31)$$

with duality with respect to Y. But $\mathcal{D}((-A^*)^\gamma) = \mathcal{D}((-A)^\gamma)$: this follows since A is the direct sum of two normal operators on Y, with possibly an additional finite-dimensional component (if 1 is an eigenvalue of A) [C-T.1], [C-T.2, Lemma A.1, case v(a) with $\alpha = \frac{1}{2}$]. Moreover, [C-T.4, with $\alpha = \frac{1}{2}$], we have

$$\mathcal{D}((-A^*)^\gamma) = \mathcal{D}((-A)^\gamma) = \mathcal{D}(A^{\frac{1}{2}+\gamma/2}) \times \mathcal{D}(A^{\gamma/2}), \quad 0 < \gamma < 1 \qquad (6.32)$$

(the first component does not really matter in the argument below). Thus, from (6.32) and B as in (6.29), it follows that (6.31) holds true, provided $\delta(x-x^0) \in [\mathcal{D}(A^{\gamma/2})]'$ (duality with respect to $L_2(\Omega)$), where $\mathcal{D}(A^{\gamma/2}) \subset H^{2\gamma}(\Omega)$, and hence, provided $\delta(x-x^0) \in [H^{2\gamma}(\Omega)]' \subset [\mathcal{D}(A^{\gamma/2})]'$. But this in turn is the case, provided $H^{2\gamma}(\Omega) \subset C(\bar{\Omega})$; i.e., by Sobolev embedding provided $2\gamma > \frac{n}{2}$, as desired. We conclude: <u>assumption</u> (1.3) $(-A)^{-\gamma}B \in \mathcal{L}(U,Y)$ <u>holds</u> <u>true</u> <u>for</u> <u>problem</u> (6.25) <u>with</u> $\frac{n}{4} < \gamma < 1$, $n \leq 3$.

<u>Assumption (H.1) = (1.5)</u>. The operator A in (6.29) generates an s.c. contraction semigroup e^{At} on Y, which moreover is analytic here for $t > 0$. (This is a special case of a much more general result

[C-T.1-2]). This, along with the requirement $\gamma < 1$ proved above guarantees that problem (6.25) satisfies assumption (H.1) = (1.5).

Remark 6.3. Since the semigroup e^{At} is analytic on Y and also uniformly stable [C-T.2], we have by the just-verified property (1.3), in the norm of $\mathcal{L}(Y,U)$:

$$\|B^* e^{A^* t}\| = \|B^* (-A^*)^{-\gamma} (-A^*)^{\gamma} e^{A^* t}\| = 0\left(\frac{1}{t^{\gamma}}\right), \quad 0 < t, \quad (6.33)$$

with $\frac{n}{4} < \gamma < 1$, $n \leq 3$. This is a sharp estimate, which for $n = 2,3$ (the interesting cases) does not allow to conclude that assumption (H.2) = (1.6) holds true. Instead, (H.2) holds true only for $n = 1$.

Finite Cost Condition (1.9). With A as in (6.29), the semigroup e^{At} is uniformly (exponentially) stable in Y [C-T.2], and thus the Finite Cost Condition (1.9) holds true with $u \equiv 0$.

Remark 6.4. Suppose that instead of Eq. (6.25a), one has

$$w_{tt} + (\Delta^2 + k_1)w - (\Delta + k_2)w_t = \delta(x - x^0)u(t) \quad \text{in } Q, \quad (6.34)$$

along with (6.25b-c). Then, if $0 < k_1 + k_2$ is sufficiently large, the generator A has finitely many unstable eigenvalues in $\{\text{Re }\lambda > 0\}$. Since e^{At} is analytic on Y, the usual theory [T.1] applies: The problem is stabilizable on Y if [T.1] and only if [M-T, Appendix] its projection onto the finite-dimensional unstable subspace is controllable.

For instance, if $\lambda_1, \ldots, \lambda_K$ are the unstable eigenvalues of A, assumed for simplicity to be simple, and $\phi_1, \ldots, \phi_K$ are the corresponding eigenfunctions in Y, then the necessary and sufficient condition for stabilization is that $\phi_k(x^0) \neq 0$, $k = 1, \ldots, K$.

If $\lambda_1, \ldots, \lambda_K$ are not simple, then their largest multiplicity M determines the smallest number of scalar controls needed for the stabilization of (6.34), where now the right hand side is replaced by $\sum_{i=1}^{M} \delta(x - x^i)u_i(t)$, along with (6.25b-c). The necessary and sufficient condition for stabilization is now a well-known full rank condition [T.1].

Detectability Condition (5.10). With R = I, this is satisfied.

Conclusion: T = ∞. Theorem 5.1 applies to problem (6.25)-(6.26), n ≤ 3, and provides existence and uniqueness of the solution to the ARE (5.1), with Riccati operator $P \in \mathcal{L}(Y, \mathcal{D}(A))$ (since A, as remarked above (6.32), is the direct sum of two normal operators on Y plus possibly a finite-dimensional component, in particular, A has a Riesz basis of eigenvectors on Y), where $\mathcal{D}(A) = \mathcal{D}(\mathcal{A}) \times \mathcal{D}(\mathcal{A}^{\frac{1}{2}})$, see (6.27), (6.28) for the characterizations of these spaces. Thus, in particular, we have $B^*P \in \mathcal{L}(Y; U)$, where $B^* \begin{bmatrix} v_1 \\ v_2 \end{bmatrix} = v_2(x^0)$. (Note that Theorem 2.2 would apply as well for n = 1.)

Conclusion: T < ∞. Theorem 5.1 applies for any G that makes GL closed (closable); see Remark 2.1, in particular for the (only) sufficient condition (2.25).

Remark 6.5. Essentially the same analysis with minimal changes applies also to problem (6.25a-b), with the B.C. (6.25c) replaced now by $\frac{\partial w}{\partial \nu}\big|_\Sigma \equiv \frac{\partial \Delta w}{\partial \nu}\big|_\Sigma \equiv 0$. The new definition of $\mathcal{A}$ incorporates, of course, these boundary conditions, and it is still true that the damping operator is precisely $\mathcal{A}^{\frac{1}{2}}$, so that A now has the same form (6.29) as before. The main difference is that the present $\mathcal{A}$ is non-negative self-adjoint and has $\mu = 0$ as an eigenvalue with corresponding one-dimensional eigenspace, spanned by the constant functions. Thus, the new operator A has $\lambda = 0$ as an eigenvalue with corresponding eigenfunction $\Phi = [\Phi_1, \Phi_2]$, $\Phi_1 = $ const, $\Phi_2 = 0$. Then, Remark 6.4 applies to stabilize the system, as the condition $\Phi(x^0) \neq 0$ is satisfied. (With no harm, one may choose to work on the space $Y = \mathcal{D}(\mathcal{A}^{\frac{1}{2}}) \times L_2^0(\Omega)$, where $L_2^0(\Omega)$ is the quotient space $L_2(\Omega)/\mathcal{N}(\mathcal{A})$, the null space of $\mathcal{A}$.)

Example 6.2: **The case** $\alpha = 1$ [C-T.1-2]. The Kelvin-Voigt model for a plate equation in the deflection w(t,x) is

$$
\begin{cases}
w_{tt} + \Delta^2 w + \rho\Delta^2 w_t = \delta(x - x^0)u(t) & \text{in } (0,T]; \Omega = Q; & (6.35a) \\[2mm]
w(0,\cdot) = w_0; \ w_t(0,\cdot) = w_1 & \text{in } \Omega; & (6.35b) \\[2mm]
\Delta w|_\Sigma + (1-\mu)B_1 w \equiv 0 & \text{in } (0,T] \times \Gamma = \Sigma; & (6.35c) \\[2mm]
\dfrac{\partial\Delta w}{\partial\nu}\Big|_\Sigma + (1-\mu)B_2 w \equiv 0 & \text{in } \Sigma; & (6.35d)
\end{cases}
$$

with $0 < \mu < \tfrac{1}{2}$ the Poisson modulus and $\rho > 0$ any constant. The boundary operators B_1 and B_2 are zero for $n = 1$, and [Lag.2] for $n = 2$:

$$
B_1 w = 2\nu_1\nu_2 w_{xy} - \nu_1^2 w_{yy} - \nu_2^2 w_{xx};
$$

$$
B_2 w = \frac{\partial}{\partial r}[(\nu_1^2 - \nu_2^2)w_{xy} + \nu_1\nu_2(w_{yy} - w_{xx})], \qquad (6.36)
$$

where again x^0 is an interior point of the open bounded $\Omega \subset R^n$, $n \leq 2$. Regularity results for problem (6.35) are given in [T.4]. Consistently with these, we take the cost functional to be the same as (6.26) with $\{w_0, w_1\} \in H^2(\Omega) \times L_2(\Omega)$.

Abstract setting. We introduce the non-negative self-adjoint operator

$$
Ah = \Delta^2 h, \ \mathcal{D}(A) = \{h \in H^4(\Omega): \Delta h + (1-\mu)B_1 h|_\Gamma = 0; \ \frac{\partial\Delta h}{\partial\nu} + (1-\mu)B_2 h|_\Gamma = 0\},
$$

$$(6.37)$$

and select the spaces and operators

$$
Y = \mathcal{D}(A^{\frac{1}{2}}) \times L_2(\Omega) = H^2(\Omega) \times L_2(\Omega); \qquad U = \mathbb{R}^1, \qquad (6.38)
$$

$$
A = \begin{vmatrix} 0 & I \\ -A & -\rho A \end{vmatrix}; \quad Bu = \begin{vmatrix} 0 \\ \delta(x-x^0)u \end{vmatrix}; \quad R = I \qquad (6.39)
$$

to obtain the abstract model (1.1), (1.2).

Assumption (1.3): $(-A)^{-\gamma}B \in \mathcal{L}(U,Y)$. Again, it is straightforward to verify that assumption (1.3) is satisfied with $\gamma = 1$: From (6.39), we require that

$$
(-A)^{-1}Bu = \begin{vmatrix} \rho I & A^{-1} \\ -I & 0 \end{vmatrix} \begin{vmatrix} 0 \\ \delta(x-x^0)u \end{vmatrix} = \begin{vmatrix} A^{-1}\delta(x-x^0)u \\ 0 \end{vmatrix} \in Y, \qquad (6.40)
$$

i.e., from (6.38) we require that $A^{-\frac{1}{4}}\delta(x-x^0)$. The same argument below (6.30) then applies yielding that (6.40) holds true if $n \leq 3$.

However, in order to verify assumption (H.1) = (1.5) which requires that γ should be < 1, the most elementary way is to check that assumption (1.3) holds in fact true with $\gamma = \frac{1}{2}$. In this case, we can in fact rely on the direct computation of $(-A)^{-\frac{1}{2}}$ (for simplicity of notation, we take henceforth $\rho = 1$)

$$(-A)^{-\frac{1}{2}} = \begin{vmatrix} (1) & A^{-\frac{3}{4}}(2I+A^{\frac{1}{2}})^{-\frac{1}{2}} \\ (2) & A^{-\frac{1}{4}}(2I+A^{\frac{1}{2}})^{-\frac{1}{2}} \end{vmatrix} \tag{6.41}$$

(where the entries (1) = $A^{-\frac{3}{4}}(2I+A^{\frac{1}{2}})^{-\frac{1}{2}}(I+A^{\frac{1}{2}})$ and (2) = $-A^{\frac{1}{4}}(2I+A^{\frac{1}{2}})^{-\frac{1}{2}}$ do not really count in the present analysis), and avoid the domain of fractional powers as in [C-T.4].

We need to compute

$$(-A)^{-\frac{1}{2}}Bu = \begin{vmatrix} A^{-\frac{3}{4}}(2I+A^{\frac{1}{2}})^{-\frac{1}{2}}\delta(x-x^0)u \\ A^{-\frac{1}{4}}(2I+A^{\frac{1}{2}})^{-\frac{1}{2}}\delta(x-x^0)u \end{vmatrix}. \tag{6.42}$$

From (6.42), we then readily see that $(-A)^{-\frac{1}{2}}Bu \in Y = \mathcal{D}(A^{\frac{1}{2}}) \times L_2(\Omega)$ provided (#): $A^{-\frac{1}{4}}\delta(x-x^0) \in L_2(\Omega)$. But $\mathcal{D}(A^{\frac{1}{2}}) = H^2(\Omega)$ (and, in fact, only $\mathcal{D}(A^{\frac{1}{4}}) \subset H^2(\Omega)$ suffices for the present analysis) so that condition (#) is satisfied provided $\delta(x-x^0) \in [H^2(\Omega)]'$ (duality with respect to $L_2(\Omega)$); i.e., provided $H^2(\Omega) \subset C(\bar{\Omega})$, i.e., by Sobolev embedding provided $2 > \frac{n}{2}$, or $n < 4$, as desired. We have shown: <u>Assumption (1.3)</u> $(-A)^{-\gamma}B \in \mathcal{L}(U,Y)$ <u>holds</u> <u>true</u> <u>for</u> <u>problem (6.35)</u> <u>with</u> $n \leq 3$, <u>and</u> $\gamma = \frac{1}{2}$. The above argument shows some 'leverage.' Indeed, $\gamma = \frac{1}{2}$ is not the least γ for which assumption (1.3) holds true. To obtain the least γ for which assumption (1.3) holds true, we proceed as in the case of problem (6.25) above, in the argument which begins with (6.31) and uses the domains of fractional powers $\mathcal{D}((-A)^{\gamma})$. As below (6.31) and ff., we need to show that (##): $Bu \in [\mathcal{D}((-A)^{\gamma})]'$, duality with respect to Y. But for $0 < \gamma \leq \frac{1}{2}$, we have from [C-T.4, with $\alpha = 1$] that

$$\mathcal{D}((-A)^{\gamma}) = \mathcal{D}(A^{\frac{1}{2}}) \times \mathcal{D}(A^{\gamma}). \tag{6.43}$$

Thus, from B as in (6.39), we see that condition (##) above holds true, provided $\delta(x-x^0) \subset [\mathcal{D}(A^{\gamma})]'$, duality with respect to $L_2(\Omega)$; i.e., provided $\delta(x-x^0) \subset [H^{4\gamma}(\Omega)]'$, since $\mathcal{D}(A^{\gamma}) \subset H^{4\gamma}(\Omega)$ for the fourth-order operator in (6.37), i.e., provided $H^{4\gamma}(\Omega) \subset C(\bar{\Omega})$, which in turn is the case, provided $4\gamma > \frac{n}{2}$, or $\frac{1}{2} \geq \gamma > \frac{n}{8}$. We conclude: Assumption (1.3) $(-A)^{-\gamma}B \in \mathcal{L}(U,Y)$ holds true for problem (6.35) provided $\frac{n}{8} < \gamma \leq \frac{1}{2}$, $n \leq 3$.

Assumption (H.1) = (1.5). The operator A in (6.39) generates an s.c. contraction semigroup e^{At} on Y, which moreover is analytic here for $t > 0$. (This is a special case of a much more general result [C-T.2].) This, along with the requirement $\gamma < 1$ proved above guarantees that problem (6.25) satisfies assumption (H.1) = (1.5).

Remark 6.6. Since the semigroup e^{At} is analytic on Y, we have by the just-verified property (1.3), in the norm of $\mathcal{L}(Y,U)$

$$\|B^* e^{A^* t}\| = \|B^*(-A^*)^{-\gamma}(-A^*)^{\gamma} e^{A^* t}\| \leq O\left(\frac{1}{t^{\gamma}}\right), \qquad 0 < t,$$

with $\frac{n}{8} < \gamma < \frac{1}{2}$, $n \leq 3$, where we can take all $t > 0$ as e^{At} is also uniformly stable [C-T.2]. Thus, for $n \leq 3$, we obtain that assumption (H.2) = (1.6) holds as well.

Finite Cost Condition (1.9). With A as in (6.39), the semigroup e^{At} is uniformly (exponentially) stable in $Y/N(A)$, the finite-dimensional nullspace of A [C-T.2], and thus the Finite Cost Condition (1.9) is automatically satisfied on this space with $u \equiv 0$. For the eigenvalue $\lambda = 0$, Remark 6.4 applies also to problem (6.35).

Detectability Condition (5.10). With R = I, this is satisfied.

Conclusion: $T = \infty$. Theorem 5.1 applies to problem (6.35) for $n \leq 3$. (Theorem 5.2 would also apply for $n \leq 3$, but the conclusions of Theorem 5.1 are stronger.)

Conclusion: $T < \infty$. Theorem 2.1 also applies to (6.35) for $n \leq 3$ for any final state operator G that makes GL closed (closable); see Remark 2.1, in particular the (only) sufficient condition (2.25).

Example 6.3. (A structurally damped plate with boundary control.) We consider the plate problem

$$w_{tt} + \Delta^2 w - \rho \Delta w_t = 0 \qquad \text{in } (0,T] \times \Omega \equiv Q, \qquad (6.44a)$$

$$w(0,\cdot) = w_0; \; w_t(0,\cdot) = w_1 \quad \text{in } \Omega, \qquad (6.44b)$$

$$w|_\Sigma \equiv 0 \qquad \text{in } (0,T] \times \Gamma \equiv \Sigma, \qquad (6.44c)$$

$$\Delta w|_\Sigma \equiv u \qquad \text{in } \Sigma, \qquad (6.44d)$$

which is the same model as the one in (6.25), except that it is acted upon by a boundary control $u \in L_2(0,T;L_2(\Gamma)) \equiv L_2(\Sigma)$, rather than a point control as in (6.25a). We take the same functional J as in (6.26) except that now u is penalized in the $L_2(\Gamma)$-norm. Following [L-T.14], we introduce the Green map G_2 defined by

$$y = G_2 v \iff \{\Delta^2 y = 0 \text{ in } \Omega; \; y|_\Gamma = 0; \; \Delta y|_\Gamma = v\}. \qquad (6.45)$$

Then, if A is the same operator defined in (6.27), it is rather straightforward to see that the abstract representation of problem (6.44) is given by the equation

$$w_{tt} + Aw + \rho A^{\frac{1}{2}} w_t = AG_2 u. \qquad (6.46)$$

(Indeed, problem (6.44) can be rewritten first as $w_{tt} + \Delta^2(w - G_2 u) - \rho \Delta w_t = 0$ in Q; $(w - G_2 v)|_\Sigma = \Delta(w - G_2 u)|_\Sigma = 0$ by (6.45); hence abstractly by $w_{tt} + A(w - G_2 u) + \rho A^{\frac{1}{2}} w_t = 0$ because of the B.C. since now $A^{\frac{1}{2}} h = -\Delta h$, $\mathcal{D}(A^{\frac{1}{2}}) = H^2(\Omega) \cap H_0^1(\Omega)$. From here, (6.46) follows by extending the original A in (6.27), as usual, by isomorphism to, say, $A: L_2(\Omega) \to [\mathcal{D}(A)]'$. It can be shown [L-T.14] that the Green map G_2 can be expressed in terms of the Dirichlet map D defined below, as follows:

$$G_2 = -A^{-\frac{1}{2}} D, \text{ where } y = Dv \iff \{\Delta y = 0 \text{ in } \Omega; \; y|_\Gamma = v\}, \qquad (6.47)$$

where D satisfies

$$D: \text{ continuous } L_2(\Gamma) \to H^{\frac{1}{2}}(\Omega) \subset H^{\frac{1}{2}-2\varepsilon}(\Omega) \equiv \mathcal{D}(A^{\frac{1}{2}-\varepsilon/2}), \quad \varepsilon > 0. \qquad (6.48)$$

Abstract setting. Thus, (6.46) becomes the abstract equation

$$w_{tt} + Aw + \rho A^{\frac{1}{2}} w_t = -A^{\frac{1}{2}} Du, \qquad (6.49)$$

or

$$\frac{d}{dt}\begin{vmatrix} w \\ w_t \end{vmatrix} = A\begin{vmatrix} w \\ w_t \end{vmatrix}; \quad A = \begin{vmatrix} 0 & I \\ -A & -\rho A^{\frac{1}{2}} \end{vmatrix}; \quad Bu = \begin{vmatrix} 0 \\ -A^{\frac{1}{2}} & Du \end{vmatrix} \tag{6.50}$$

on the spaces $Y = \mathcal{D}(A^{\frac{1}{2}}) \times L_2(\Omega)$; $U = L_2(\Gamma)$.

Assumption (1.3). $(-A)^{-\gamma}B \in \mathcal{L}(U,Y)$. Again, it is elementary to verify that assumption (1.3) is satisfied with $\gamma = 1$: Indeed, from (6.50) we require

$$(-A)^{-1}Bu = \begin{vmatrix} \rho A^{-\frac{1}{2}} & A^{-1} \\ -I & 0 \end{vmatrix} \begin{vmatrix} 0 \\ -A^{\frac{1}{2}} & Du \end{vmatrix} = \begin{vmatrix} -A^{-\frac{1}{2}} & Du \\ 0 \end{vmatrix} \in Y = \mathcal{D}(A^{\frac{1}{2}}) \times L_2(\Omega), \tag{6.51}$$

which certainly holds true by (6.48). We may also verify that the value $\gamma = \frac{1}{2}$ fails: from direct computations (as in (6.41)) or from [T.4], we obtain (say with $\rho = 1$)

$$(-A)^{-\frac{1}{2}}Bu = \frac{-1}{\sqrt{3}}\begin{vmatrix} A^{-\frac{3}{4}} & A^{\frac{1}{2}}Du \\ A^{-\frac{1}{4}} & A^{\frac{1}{2}}Du \end{vmatrix} = \frac{-1}{\sqrt{3}}\begin{vmatrix} A^{-\frac{1}{4}}Du \\ A^{\frac{1}{4}}Du \end{vmatrix}, \tag{6.52}$$

and from (6.48) we see that $(-A)^{-\frac{1}{2}}Bu$ in (6.52) fails by $\frac{1}{4} + \varepsilon/2$, to be in Y.

Indeed, we have that: <u>Assumption (1.3) holds true for all</u> $\frac{3}{4} < \gamma < 1$.

The above claim can be verified by an argument similar to the ones of the preceding two examples, based on the domain of fractional power [C-T.4, with $\alpha = \frac{1}{2}$]

$$\mathcal{D}((-A)^{\gamma}) = \mathcal{D}(A^{\frac{1}{2}+\gamma/2}) \times \mathcal{D}(A^{\gamma/2}), \quad \frac{1}{2} \leq \gamma < 1 \tag{6.53}$$

(only the second component is needed in our argument), whereby the usual condition $Bu \in [\mathcal{D}((-A)^{\gamma})]'$, duality with respect to Y, is satisfied provided, from (6.50) and (6.53), $A^{\frac{1}{2}}Du \in [\mathcal{D}(A^{\gamma/2})]'$, duality with respect to $L_2(\Omega)$; i.e., from (6.48) provided $\frac{3}{8} + \frac{\varepsilon}{2} \leq \frac{\gamma}{2}$ or $\gamma = \frac{3}{4} + \varepsilon$, $\forall \varepsilon$, and the claim is proved.

Assumption (H.1) = (1.5). This is satisfied as $\gamma < 1$ was proved above and e^{At} is an s.c. contraction, analytic semigroup on Y [C-T.1-2].

Remark 6.7. Since $\frac{1}{2} < \gamma$, a computation as in (6.33) shows that assumption (H.2) = (1.6) is not satisfied.

Finite Cost Condition (1.9). With A as in (6.50), the semigroup e^{At} is uniformly (exponentially) stable on Y [C-T.2], and thus the Finite Cost Condition (1.9) is automatically satisfied with u $\equiv$ 0 (see also Remark 3.4).

Conclusion: T = ∞. Theorem 5.1 with R = I applies to problem (6.44) for any n. (Theorem 2.2 is not applicable.)

Conclusion: T < ∞. Theorem 2.1 applies with any G such that GL is closed (closable); see Remark 2.1, in particular the (only) sufficient condition (2.25).

Remark 6.8. A similar analysis applies if the B.C. (6.44c-d) are replaced by $\frac{\partial w}{\partial \nu}\big|_\Sigma \equiv 0$, $\frac{\partial \Delta w}{\partial \nu}\big|_\Sigma = u$; refer to Remark 6.5.

So far, we have considered examples of damped plates where the damping operator is equal to the α-th power of the original elastic differential operator, $\alpha = \frac{1}{2}$ in Examples 6.1 and 6.3, and $\alpha = 1$ in Example 6.2. This was due to the special choice of boundary conditions. In the next example, we return to the same Eq. (6.25a), which we now complement with boundary conditions that make the damping operator only 'comparable,' in a technical sense [C-T.1-2], to the α-th power of the elastic operator, $\alpha = \frac{1}{2}$.

Example 6.5. On some smooth Ω, dim Ω = n $\leq$ 3, consider the plate model with $\rho > 0$ any constant:

$$\begin{cases} w_{tt} + \Delta^2 w - \rho \Delta w_t = \delta(x - x^0) u(t) & \text{in } (0,T] \times \Omega = Q, & (6.54a) \\ w(0,\cdot) = w_0; \ w_t(0,\cdot) = w_1 & \text{in } \Omega, & (6.54b) \\ w\big|_\Sigma \equiv \frac{\partial w}{\partial \nu}\big|_\Sigma \equiv 0 & \text{in } (0,T] \times \Gamma \equiv \Sigma, & (6.54c) \end{cases}$$

Regularity results for problem (6.54) are given in [T.4]. Consistently with these, we associate with (6.54) the same cost functional (6.26).

Abstract setting. Now, however, we introduce the positive self-adjoint operators

$$Ah = \Delta^2 h, \quad \mathcal{D}(A) = \{h \in H^4(\Omega): h|_\Gamma = \frac{\partial h}{\partial \nu}|_\Gamma = 0\} \qquad (6.55)$$

$$\mathcal{B}h = -\Delta h, \quad \mathcal{D}(\mathcal{B}) = \{h \in H^2(\Omega): h|_\Gamma = \frac{\partial h}{\partial \nu}|_\Gamma = 0\} = H_0^2(\Omega) = \mathcal{D}(A^{\frac{1}{2}}), \quad (6.56)$$

where the equality with $\mathcal{D}(A^{\frac{1}{2}})$ in (6.26) (equivalent norms) is standard [Gr]. Thus, problem (6.54) admits the abstract version

$$w_{tt} + Aw + \rho \mathcal{B} w_t = \delta(x-x^0)u$$

which fits the abstract model (1.1), (1.2) with

$$A = \begin{vmatrix} 0 & I \\ -A & -\rho\mathcal{B} \end{vmatrix}; \quad Bu = \begin{vmatrix} 0 \\ \delta(x-x^0)u \end{vmatrix}; \quad R = I \qquad (6.57)$$

on the spaces $Y = \mathcal{D}(A^{\frac{1}{2}}) \times L_2(\Omega)$, $U = \mathbb{R}^1$. From (6.56), we have

$$\mathcal{B}^2 h = \Delta^2 h, \quad \mathcal{D}(\mathcal{B}^2) = \{h \in H^4(\Omega): h|_\Gamma = \frac{\partial h}{\partial \nu}|_\Gamma = \Delta h|_\Gamma = \frac{\partial \Delta h}{\partial \nu}|_\Gamma = 0\} \subset \mathcal{D}(A).$$
$$(6.58)$$

By Green's second theorem, we have

$$(\mathcal{B}^2 f, f) = (\Delta^2 f, f) = \int_\Gamma \frac{\partial \Delta f}{\partial \nu} f \, d\Gamma - \int_\Gamma \Delta f \frac{\partial f}{\partial \nu} \, d\Gamma + \int_\Omega (\Delta f)^2 d\Gamma \qquad (6.59)$$

$$= \int_\Omega (\Delta f)^2 d\Omega, \quad f \in \mathcal{D}(\mathcal{B}^2); \qquad (6.60)$$

$$(Af, f) = (\Delta^2 f, f) = \int_\Gamma \frac{\partial \Delta f}{\partial \nu} f \, d\Gamma - \int_\Gamma \Delta f \frac{\partial f}{\partial \nu} \, d\Gamma + \int_\Omega (\Delta f)^2 d\Omega \qquad (6.61)$$

$$= \int_\Omega (\Delta f)^2 d\Omega, \quad f \in \mathcal{D}(A); \qquad (6.62)$$

where the boundary terms on the right hand side of (6.59) and (6.61) still vanish if f is only in $\mathcal{D}(\mathcal{B})$, or $\mathcal{D}(A^{\frac{1}{2}})$, respectively, see (6.56). Thus, by extension, we get

$$(\mathcal{B}^2 f, f) = (Af, f) = \int_\Omega (\Delta f)^2 d\Omega, \quad f \in \mathcal{D}(\mathcal{B}) = \mathcal{D}(A^{\frac{1}{2}}), \qquad (6.63)$$

and thus *a fortiori* the results of [C-T.1-2] apply: These in particular establish that the operator A in (6.57) <u>generates an s.c.</u>, <u>analytic semigroup</u> e^{At} <u>on</u> Y.

Assumption (1.3): $(-A)^{-\gamma}B \in \mathcal{L}(U,Y)$. Again, it is immediate to see that assumption (1.3) holds true for $\gamma = 1$. In fact, as in the argument below (6.30), we find that

$$(-A)^{-1}Bu = \begin{vmatrix} A^{-1}\mathcal{B} & A^{-1} \\ -I & 0 \end{vmatrix} \begin{vmatrix} 0 \\ \delta(x-x^0) \end{vmatrix} = \begin{vmatrix} A^{-1}\delta(x-x^0) \\ 0 \end{vmatrix} \in Y. \quad (6.64)$$

In effect, _assumption (1.3) holds true for problem_ (6.54) _for all_ γ _with_ $\frac{n}{4} < \gamma < 1$, $n \leq 3$, exactly as in the case of problem (6.1). To see this, with A and $\mathcal{B}$ as in (6.55), (6.56), we denote for convenience

$$A_{\mathcal{B}} = \begin{vmatrix} 0 & I \\ -A & -\mathcal{B} \end{vmatrix}; \qquad A_{\mathcal{B}}^{*} = \begin{vmatrix} 0 & -I \\ A & -\mathcal{B}^{*} \end{vmatrix};$$

$$A_{\frac{1}{2}} = \begin{vmatrix} 0 & I \\ -A & -A^{\frac{1}{2}} \end{vmatrix}; \qquad A_{\frac{1}{2}}^{*} = \begin{vmatrix} 0 & -I \\ A & -A^{\frac{1}{2}} \end{vmatrix}; \qquad (6.65)$$

where the adjoints are with respect to Y. Since $\mathcal{B}$ in (6.56) is self-adjoint on $L_2(\Omega)$, we have

$$\mathcal{D}(A_{\mathcal{B}}^{*}) = \mathcal{D}(A_{\mathcal{B}}) = \mathcal{D}(A_{\frac{1}{2}}) = \mathcal{D}(A_{\frac{1}{2}}^{*}). \quad (6.66)$$

As a consequence of (part of) (6.58), we have in our present case [C-T.4]

$$\mathcal{D}((-A_{\frac{1}{2}})^{\gamma+\varepsilon}) \subset \mathcal{D}((-A_{\mathcal{B}}^{*})^{\gamma}) \subset \mathcal{D}((-A_{\frac{1}{2}})^{\gamma-\varepsilon}), \quad 0 < \gamma < 1, \quad (6.67)$$

and $\gamma+\varepsilon < 1$. Then, to obtain $(-A_{\mathcal{B}})^{-\gamma}B \in \mathcal{L}(U,Y)$, $\frac{n}{4} < \gamma < 1$, as desired, i.e., $Bu \in [\mathcal{D}((-A_{\mathcal{B}}^{*})^{\gamma}]'$, duality with respect to Y, it suffices via the right hand side of (6.67) to have $Bu \in [\mathcal{D}((-A_{\frac{1}{2}})^{\gamma-\varepsilon})]'$. But this was shown to be true in Example 6.1, precisely for $\frac{n}{4} < \gamma-\varepsilon < 1$, in the argument below (6.31).

Alternatively, we may write, using Example 6.1, that

$$(-A_{\mathcal{B}})^{-\gamma}B = (-A_{\mathcal{B}})^{-\gamma}(-A_{\frac{1}{2}})^{\gamma-\varepsilon}(-A_{\frac{1}{2}})^{-(\gamma-\varepsilon)}B \in \mathcal{L}(U,Y), \quad (6.68)$$

since $(-A_{\mathcal{B}})^{-\gamma}(-A_{\frac{1}{2}})^{\gamma-\varepsilon}$ is bounded, for $(-A_{\frac{1}{2}}^{*})^{\gamma-\varepsilon}(-A_{\mathcal{B}}^{*})^{-\gamma}$ is bounded by (6.66) and the closed graph theorem.

Assumption (H.1) = (1.5). This assumption holds true for problem (6.54) since it was already observed below (6.63) that e^{At} is a s.c. analytic, semigroup on Y, while γ was shown above to be < 1 for $n \leq 3$.

The Finite Cost Condition (1.9) and **The Detectability Condition (5.10)**. These also hold true, since e^{At} is uniformly stable [C-T.2] and R = I.

Conclusion: $T = \infty$. Theorem 5.1 applies to problem (6.54).

Conclusion: $T < \infty$. Theorem 2.1 applies to problem (6.54) with any G such that GL is closed (closable); see Remark 2.1, in particular the (only) sufficient condition (2.25).

Example 6.6. (The case $\alpha = 1$ in [C-T.1-2]) On some smooth bounded $\Omega \subset R^n$ we consider the wave equation with a strong degree of damping and $\rho > 0$:

$$\begin{cases} w_{tt} - \Delta w - \rho \Delta w_t = \delta(x - x^0)u(t) & \text{in } (0,T] \times \Omega = Q; & (6.69a) \\ w(0, \cdot) = w_0; \; w_t(0, \cdot) = w_1 & \text{in } \Omega; & (6.69b) \\ w|_\Sigma \equiv 0 & \text{in } (0,T] \times \Gamma = \Sigma. & (6.69c) \end{cases}$$

Problem (6.69) can be written abstractly as

$$w_{tt} + Aw + \rho A w_t = \delta(\cdot - x^0)u(t), \qquad (6.70)$$

i.e., precisely as problem (6.35), except that now A is

$$A = -\Delta, \quad \mathcal{D}(A) = H^2(\Omega) \cap H_0^1(\Omega). \qquad (6.71)$$

Case 1. We begin by selecting

$$Y = \mathcal{D}(A^{1/2}) \times L_2(\Omega) = H_0^1(\Omega) \times L_2(\Omega); \quad U = R^1. \qquad (6.72)$$

The cost functional for $T = \infty$ is then

$$J(u,w) = \int_0^\infty \{\|w(t)\|_{H^1(\Omega)}^2 + \|w_t(t)\|_{L_2(\Omega)}^2 + |u(t)|^2\} dt. \qquad (6.73)$$

Assumption (1.3): $(-A)^{-\gamma}B \in \mathcal{L}(U,Y)$. The case $\gamma = 1$ follows as in (6.40), since A and B are the same expressions as in (6.39): we require

$A^{-1}\delta \in \mathcal{D}(A^{\frac{1}{2}})$, i.e., $A^{-\frac{1}{2}}\delta \in L_2(\Omega)$, i.e., $\delta \in H^{-1}(\Omega)$. And this is true only for dim $\Omega = n = 1$. Indeed $\gamma = \frac{1}{2}$ works as in (6.41), (6.42): we have $(-A)^{-\frac{1}{2}}Bu \in Y$ provided $A^{-\frac{1}{2}}\delta \in L_2(\Omega)$, i.e., for $n = 1$.

Assumption (H.1) = (1.5). Now A generates a contraction analytic semigroup e^{At} on Y [C-T.2] and $\gamma = \frac{1}{2} < 1$ was shown above.

To remove the limitation $n = 1$ of Case 1, we now consider

Case 2. We choose now the following spaces and cost functional for $T = \infty$:

$$Y = L_2(\Omega) \times L_2(\Omega); \quad U = R^1; \quad (6.74)$$

$$J(u,w) = \int_0^\infty \{\|w(t)\|^2_{L_2(\Omega)} + \|w_t(t)\|^2_{L_2(\Omega)} + |u(t)|^2\}dt. \quad (6.75)$$

Assumption (1.3): $(-A)^{-\gamma}B \in \mathcal{L}(U,Y)$. As before, using (6.40), we see that the case $\gamma = 1$ requires now that $A^{-1}\delta \in \mathcal{L}_2(\Omega)$. Since $\mathcal{D}(A) \subset H^2(\Omega)$, this requirement is then satisfied provided $\delta \in [H^2(\Omega)]'$, which in turn is true provided $H^2(\Omega) \subset C(\bar{\Omega})$, i.e., provided $2 > \frac{n}{2}$ or $n < 4$. Indeed, we shall show that: (#) $(-A)^{-1}Bu \in \mathcal{D}((-A)^\Theta)$, $u \in R^1$, $\Theta < \frac{1}{4}$, $n \le 3$, so that condition (1.3) holds true for any $\gamma < \frac{3}{4}$ with $n \le 3$. Indeed, we have that $\mathcal{D}((-A)^\Theta) = \{x, y \in L_2(\Omega): x + \rho y \in \mathcal{D}(A^\Theta)\}$. This result follows by adapting the proof of [C-T.4] with $\alpha = 1$ to A defined on $L_2(\Omega) \times L_2(\Omega)$ as in (6.74) rather than on $\mathcal{D}(A^{\frac{1}{2}}) \times L_2(\Omega)$ as in [C-T.4]. Consequently, recalling (6.40), we see that condition (#) holds true if and only if $A^{-1}\delta \in \mathcal{D}(A^\Theta)$, or $A^{-(1-\Theta)}\delta \in L_2(\Omega)$. Since $\mathcal{D}(A^{1-\Theta}) \subset H^{2-2\Theta}(\Omega)$, we see that this latter condition holds true in case $\delta \in [H^{2-2\Theta}(\Omega)]'$, i.e., provided $H^{2-2\Theta}(\Omega) \subset C(\bar{\Omega})$, i.e., provided $2-2\Theta > \frac{n}{2}$. For $\Theta < \frac{1}{4}$ we have $4-4\Theta > 3 \ge n$, as desired.

Assumption (H.1) = (1.5). The operator A generates a s.c., analytic semigroup e^{At} on Y (not contraction now) since, as one sees readily, $\|R(\lambda,A)\| \le C/|\lambda|$ for Re $\lambda > 0$ in the norm of $\mathcal{L}(Y)$ [C-T.2]. That $\gamma < 1$ was shown above for $n \le 3$.

Finite Cost Condition (1.9). The semigroup e^{At} is uniformly stable with either choice of Y, (6.72) or (6.74).

Conclusion: $T = \infty$. Theorem 5.1 applies to the cost (6.73) for $n = 1$ and to the cost (6.75) for $n \leq 3$.

Conclusion: $T < \infty$. Theorem 2.1 applies with Y as in (6.72) for $n = 1$ and to Y as in (6.74) for $n \leq 3$ for any G that makes GL closed (closable); see Remark 2.1 in particular the (only) sufficient condition (2.25).

7. **Examples of partial differential equation problems satisfying (H.2)**

In this section, we illustrate the applicability of Theorem 5.2 within the class of dynamics subject to hypothesis (H.2) = (1.6). Our examples include: the wave equation with Dirichlet control (Section 7.1); the Euler-Bernoulli equation with Dirichlet/Neumann controls; and with controls as displacement/bending moment (Section 7.2); the Schrödinger equation with Dirichlet control (Section 7.4); first-order hyperbolic systems (Section 7.5); finally, Kirchoff plate with boundary control as a 'bending moment' (Section 7.6).

7.1. **Class (H.2): Second order hyperbolic equations with Dirichlet boundary control**

We consider the following problem:

$$\begin{cases} w_{tt} = \Delta w & \text{in } (0,T] \times \Omega = Q, & (7.1a) \\ w(0,\cdot) = w_0, \ w_t(0,\cdot) = w_1 & \text{in } \Omega, & (7.1b) \\ w|_\Sigma \equiv u & \text{in } (0,T] \times \Gamma \equiv \Sigma, & (7.1c) \end{cases}$$

where we take the boundary control $u \in L_2(\Sigma)$. (In (7.1a) we may replace $-\Delta$ by any second order, elliptic operator with time independent, symmetric coefficients of its principal part, with minimal changes in the analysis below.) Consistently with established (optimal) regularity theory [Lio.1], [L-T.2], [LLT], we take $\{w_0, w_1\} \in L_2(\Omega) \times H^{-1}(\Omega)$ and the cost functional

$$J(u,w) = \int_0^\infty \{\|w(t)\|^2_{L_2(\Omega)} + \|w_t(t)\|^2_{H^{-1}(\Omega)} + |u(t)|^2_{L_2(\Gamma)}\}dt. \quad (7.2)$$

Abstract setting [T.2], [L-T.1], [L-T.2]. To put problem (7.1), (7.2) into the abstract model (1.1), (1.2), we introduce the positive self-adjoint operator $Ah = -\Delta h$, $\mathcal{D}(A) = H^2(\Omega) \cap H^1_0(\Omega)$ and define the operators

$$A = \begin{vmatrix} 0 & I \\ -A & 0 \end{vmatrix}; \quad Bu = \begin{vmatrix} 0 \\ ADu \end{vmatrix}; \quad R = I, \tag{7.3}$$

where D is the Dirichlet map encountered before in Example 6.3, Eq. (6.47),

$$Dv = y \iff \{\Delta y = 0 \text{ in } \Omega, \ y|_\Gamma = v\}, \tag{7.4}$$

and the spaces

$$Z = Y = L_2(\Omega) \times H^{-1}(\Omega); \quad U = L_2(\Gamma). \tag{7.5}$$

The Dirichlet map satisfies the regularity property (6.48).

Assumption (1.3): $(-A)^{-\gamma}B \in \mathcal{L}(U,Y)$. From (7.3) with $u \in L_2(\Gamma)$, we obtain

$$(-A)^{-1}Bu = \begin{vmatrix} 0 & A^{-1} \\ -I & 0 \end{vmatrix} \begin{vmatrix} 0 \\ ADu \end{vmatrix} = \begin{vmatrix} Du \\ 0 \end{vmatrix} \in Y, \tag{7.6}$$

a fortiori, by the regularity (6.48) of D, and assumption (1.3) holds true with $\gamma = 1$.

Assumption (H.2) = (1.6). From (7.3) we calculate

$$B^* \begin{vmatrix} z_1 \\ z_2 \end{vmatrix} = D^* z_2 = -\frac{\partial}{\partial \nu} A^{-1} z_2, \quad \text{since } D^* A = -\frac{\partial}{\partial \nu} \text{ [L-T.2]}. \tag{7.7}$$

Moreover, we have

$$B^* e^{A^* t} \begin{vmatrix} z_1 \\ z_2 \end{vmatrix} = \frac{\partial \phi(t)}{\partial \nu}, \quad [z_1, z_2] \in Y, \tag{7.8}$$

where $\phi(t) = \phi(t, \phi_0, \phi_1)$ solves the corresponding homogeneous problem

$$\begin{cases} \phi_{tt} = \Delta\phi & \text{in } (0,T] \times \Omega \equiv Q, & (7.9a) \\ \phi(0, \cdot) = \phi_0, \ \phi_t(0, \cdot) = \phi_1 & \text{in } \Omega, & (7.9b) \\ \phi|_\Sigma \equiv 0 & \text{in } (0,T] \times \Gamma \equiv \Sigma, & (7.9c) \end{cases}$$

with

$$\phi_0 = -A^{-1}z_2 \in \mathcal{D}(A^{\frac{1}{2}}) = H_0^1(\Omega); \quad \phi_1 = z_1 \in L_2(\Omega). \qquad (7.10)$$

Thus, by (7.8), (7.10), an equivalent formulation of assumption (H.2) = (1.6) is the inequality

$$\int_\Sigma \left(\frac{\partial\phi}{\partial\nu}\right)^2 d\Sigma \leq c_T \|\{\phi_0, \phi_1\}\|^2_{H_0^1(\Omega)\times L_2(\Omega)} \qquad (7.11)$$

for the trace of the solution to problem (7.9). It should be noted that inequality (7.11) does NOT follow from a priori (optimal) interior regularity $\phi(t) \in C([0,T];H_0^1(\Omega))$ of the solution to problem (7.9), (7.10). Inequality (7.11) is an independent trace regularity result. It was first established in [L-T.1], [L-T.2]: In these references it was first proved, by means of pseudo-differential operator techniques, that the following interior regularity result holds true: $\{w, w_t\} \in L_2(0,T;L_2(\Omega)\times H^{-1}(\Omega))$ for problem (7.1) with $u \in L_2(\Sigma)$, $\{w_0, w_1\} \in L_2(\Omega)\times H^{-1}(\Omega)$. Next, via a duality argument, it was proved by a purely operator technique, that indeed (7.11) holds true, and that, in fact, $\{w, w_t\} \in C([0,T]; L_2(\Omega)\times H^{-1}(\Omega))$. Inequality (7.11) was proved independently and directly also in [Lio.1], by a multiplier technique; see also [LLT] for a comprehensive treatment.

Finite Cost Condition (1.9). Sufficient conditions which would imply that the F.C.C. (1.9) is satisfied are:

(i)   (exponential) uniform stabilization of problem (7.1) on the space $Y = L_2(\Omega)\times H^{-1}(\Omega)$ by means of an $L_2(0,\infty;L_2(\Gamma))-$ feedback u;

(ii)   exact controllability of problem (7.1) (to or, equivalently, from the origin) over a finite interval [0,T], on the state space $Y = L_2(\Omega)\times H^{-1}(\Omega)$, within the class of $L_2(0,T;L_2(\Gamma))-$ controls u.

A solution to the uniform stabilization problem (i), and consequently, via a known result of D. Russell (1973) of the exact controllability problem (ii) was first obtained in [L-T.12], under some additional geometrical condition on Ω (which includes the class of strictly convex Ω). Later, exact controllability, this time without geometrical conditions on Ω, except for smooth Γ, if u is applied to all of Γ, was established in [Lio.2] by relying on a lower bound

inequality, inequality (7.11) with the reversed inequality sign, if T
is sufficiently large (twice the diameter of Ω)

$$\int_\Sigma \left(\frac{\partial\phi}{\partial\nu}\right)^2 d\Sigma \geq c_T \|\{\phi_0,\phi_1\}\|^2_{H^1_0(\Omega)\times L_2(\Omega)}. \tag{7.12}$$

This latter inequality (7.12) was explicitly obtained in [H]--by using
the same multiplier methods that had been used in [Lio.1], [L-L-T] for
inequality (7.11)--and in [L-T.12] to solve the uniform stabilization
problem; indeed, such inequality (7.12) is essentially contained also
in the estimates of this work [L-T.12], albeit in a less transparent
form. A direct approach to exact controllability based on the
surjectivity of the input-solution operator and multiplier methods to
show the key inequality, in the case where u acts only on a portion of
the boundary Σ_0 is given in [T.3]. Later, geometric optics methods--
first introduced in [Lit.] for exact controllability questions--
provided sharp sufficient conditions for inequality (7.12) to hold
true, with Σ replaced by a subportion $\Sigma_0 \subset \Sigma$ [B-L-R]. The uniform
stabilization problem now holds with no geometrical conditions in Ω
[L-T.25], if the feedback acts on all of Γ. In any case, the validity
of the Finite Cost Condition (1.9) for problem (7.1), or a more general
version thereof, is firmly established.

        We note, however, that if the control u in (7.1c) is sought
within the class of finitely many actuators

$$u(t,x) = \sum_{j=1}^{J} g_j(x)\mu_j(t)$$

with $J < \infty$, $g_j \in L_2(\Gamma)$ arbitrary but fixed, and $\mu_j \in L_2(0,T)$, then exact
controllability on any [0,T] within the class of μ_j-controls is not
possible for problem (7.1) on the required space Y in (7.5), unless
dim Ω = 1 [T.8]. This comment applies to all other cases in this
section [T.8] and will not be repeated.

Detectability Condition (5.17)-(5.19). This holds true since R = I.

        However, we find instructive to give a non-trivial example for
the wave dynamics (7.1) with penalization in $L_2(\Omega)\times H^{-1}(\Omega)$ involving an

observation operator R which is not positive definite. Instead of the cost function (7.2) we consider now

$$J(u,w) = \int_0^\infty \{\|w(t)\|^2_{L_2(\Omega)} + \|mw_t(t)\|^2_{H^{-1}(\Omega)} + |u(t)|^2_{L_2(\Gamma)}\}dt ,$$

where m(x) is a smooth non-negative function defined on Ω, with compact support on a proper subdomain Ω_0 of Ω, so that the new functional picks up w_t only on Ω_0. Define $R = diag[R_1, R_2]$, $R_1 = 0$, $(R_2f)(x) = m(x)f(x)$, $f \in H^{-1}(\Omega)$, so that R_2 is a (self-adjoint) multiplication operator. In order to satisfy the Detectability Condition (5.17)-(5.19), we take $K = diag[0,-I]$, so that the feedback problem corresponding to (5.18) is now $\dot{y} = (A+KR)y$, or the abstract equation $w_{tt} = -Aw-R_2w_t$; i.e., the p.d.e. problem with 'viscous damping'

$$\begin{cases} w_{tt} = \Delta w - mw_t & \text{in } Q; \\ w(0,\cdot) = w_0, \ w_t(0,\cdot) = w_1 & \text{in } \Omega; \\ w|_\Sigma \equiv 0 & \text{in } \Sigma. \end{cases}$$

The Detectability Condition requires that the y-problem, equivalently the $\{w,w_t\}$-problem, be uniformly stable in the topology of $L_2(\Omega) \times H^{-1}(\Omega)$. And this is the case if and only if all rays of geometric optics meet the set $\Omega \times [0,T]$ [B-L-R].

Conclusion: $T = \infty$. Theorem 5.2 applies to problem (7.1) in the two cases described.

Conclusion: $T < \infty$. We have already noted that Theorem 3.1 holds true for the wave equation problem (7.1), (7.2), where R = Identity, while Theorem 3.2 to be applicable for (7.1) requires some minimal smoothing on R as described there. Finally, Theorem 3.3 applies for (7.1) with additional smoothing on R as described in Remark 3.4.

7.2. **Class (H.2): Euler-Bernoulli equations with boundary control**

Case 1. We consider on any smooth bounded $\Omega \subset R^n$:

$$\begin{cases} w_{tt} + \Delta^2 w = 0 & \text{in } (0,T] \times \Omega = Q, & (7.13a) \\[1mm] w(0,\cdot) = w_0, \ w_t(0,\cdot) = w_1 & \text{in } \Omega, & (7.13b) \\[1mm] w|_\Sigma \equiv 0 & \text{in } (0,T] \times \Gamma = \Sigma, & (7.13c) \\[1mm] \dfrac{\partial w}{\partial \nu}\Big|_\Sigma \equiv u & \text{in } \Sigma, & (7.13d) \end{cases}$$

with boundary control $u \in L_2(\Sigma)$. Consistently with optimal regularity theory, the cost functional to be minimized is

$$J(u,w) = \int_0^\infty \|w(t)\|^2_{L_2(\Omega)} + \|w_t(t)\|^2_{H^{-2}(\Omega)} + \|u(t)\|^2_{L_2(\Gamma)} dt. \quad (7.14)$$

Abstract setting. To put problem (7.13), (7.14) into the abstract model (1.1), (1.2), we introduce the positive self-adjoint operator

$$Ah = \Delta^2 h, \quad \mathcal{D}(A) = \{h \in H^4(\Omega) : h|_\Gamma = \frac{\partial h}{\partial \nu}\Big|_\Gamma = 0\}. \quad (7.15)$$

and define the operators

$$A = \begin{vmatrix} 0 & I \\ -A & 0 \end{vmatrix}; \quad Bu = \begin{vmatrix} 0 \\ AG_2 u \end{vmatrix}; \quad R = I \quad (7.16)$$

where G_2 is the appropriate Green map:

$$y = G_2 v \iff \{\Delta^2 y = 0 \text{ in } \Omega; \ y|_\Gamma = 0, \ \frac{\partial y}{\partial \nu}\Big|_\Gamma = v\}, \quad (7.17)$$

and the spaces

$$Y \equiv L_2(\Omega) \times H^{-2}(\Omega), \quad U \equiv L_2(\Gamma). \quad (7.18)$$

Assumption (1.3): $(-A)^{-\gamma} B \in \mathcal{L}(U,Y)$. Since G_2 is certainly bounded $L_2(\Gamma) \to L_2(\Omega)$, we readily obtain from (7.16) with $u \in L_2(\Gamma)$:

$$(-A)^{-1} Bu = \begin{vmatrix} 0 & A^{-1} \\ -I & 0 \end{vmatrix} \begin{vmatrix} 0 \\ AG_2 u \end{vmatrix} = \begin{vmatrix} G_2 u \\ 0 \end{vmatrix} \in Y, \quad (7.19)$$

and assumption (1.3) holds true for problem (7.13) with $\gamma = 1$.

Assumption (H.2) = (1.6). One can show that [L-T.9], [FLT, Appendix C],

$$B^* e^{A^* t} \begin{vmatrix} y_1 \\ y_2 \end{vmatrix} = \Delta\phi(t)\big|_\Gamma, \qquad y \in Y, \qquad (7.20)$$

where $\phi(t) = \phi(t, \phi_0, \phi_1)$ solves the corresponding homogeneous problem

$$\begin{cases} \phi_{tt} + \Delta^2\phi = 0 & \text{in } (0,T]\times\Omega = Q, & (7.21a) \\ \phi(0,\cdot) = \phi_0, \ \phi_t(0,\cdot) = \phi_1 & \text{in } \Omega, & (7.21b) \\ \phi\big|_\Sigma \equiv \dfrac{\partial\phi}{\partial\nu}\big|_\Sigma \equiv 0 & \text{in } (0,T]\times\Gamma \equiv \Sigma, & (7.21c) \end{cases}$$

with

$$\phi_0 = -A^{-1}y_2 \in \mathcal{D}(A^{\frac{1}{2}}) = H_0^2(\Omega); \ \phi_1 = y_1 \in L_2(\Omega). \qquad (7.22)$$

Thus, by (7.20), (7.22), an equivalent formulation of assumption (H.2) = (1.6) is the inequality

$$\int_\Sigma |\Delta\phi|^2 d\Sigma \le c_T \|\{\phi_0, \phi_1\}\|^2_{H_0^2(\Omega)\times L_2(\Omega)} \qquad (7.23)$$

for the trace of the solution to problem (7.21). As in the case of the wave equation of section 7.1, it should be noted that inequality (7.23) does NOT follow from *a priori* (optimal) interior regularity $\phi(t) \in C([0,T];H_0^2(\Omega))$ of the solution to the problem (7.21), (7.22). It is an independent regularity result which holds indeed true [Lio.2], for any general smooth Ω. Thus, assumption (H.2) = (1.6) holds true for problem (7.13).

Finite Cost Condition (1.9). The same considerations apply now as in the case of the preceding wave equation problem. Exact controllability of problem (7.13) holds true, for any T > 0, in fact, on the state space $Y = L_2(\Omega)\times H^{-2}(\Omega)$ within the class of $L_2(\Sigma)$-controls u, with no geometrical conditions on Ω [Lio.2], [L-T.28] if the control acts on all of Γ. Uniform stabilization can also be established, likewise without geometrical conditions [O-T].

Detectability Conditions (5.17)-(5.19). This holds true since R = I.

Conclusion: T = ∞. Theorem 5.2 applies to problem (7.13), (7.14).

Conclusion: T < ∞: Theorem 3.1 applies to (7.13), (7.14) where R = I; while Theorem 3.3 requires a stronger smoothing assumption on R.

__Case 2__. We now consider on any smooth $\Omega \subset R^n$:

$$
\begin{cases}
w_{tt} + \Delta^2 w = 0 & \text{in } Q, & (7.24a) \\[4pt]
w(0, \cdot) = w_0, \ w_t(0, \cdot) = w_1 & \text{in } \Omega, & (7.24b) \\[4pt]
w|_\Sigma = u & \text{in } \Sigma, & (7.24c) \\[4pt]
\frac{\partial w}{\partial \nu}\Big|_\Sigma = 0 & \text{in } \Sigma, & (7.24d)
\end{cases}
$$

with boundary control $u \in L_2(\Sigma)$. Consistently with optimal regularity theory [Lio.2], [L-T.11], the cost functional to be minimized is

$$
J(u,w) = \int_0^\infty \{\|w(t)\|^2_{H^{-1}(\Omega)} + \|w_t(t)\|^2_{V'} + \|u(t)\|^2_{L_2(\Gamma)}\}dt. \qquad (7.25)
$$

where V' is the dual space of V defined by

$$
V = \{h \in H^3(\Omega): \ h|_\Gamma = \frac{\partial h}{\partial \nu}\Big|_\Gamma = 0\}. \qquad (7.26)
$$

__Abstract setting__. To put problem (7.24), (7.25) into the abstract model (1.1), (1.2), we introduce the operators

$$
A = \begin{vmatrix} 0 & I \\ -A & 0 \end{vmatrix}; \quad Bu = \begin{vmatrix} 0 \\ AG_1 u \end{vmatrix}; \quad R = I \qquad (7.27)
$$

with A the operator in (7.15) and G_1 the appropriate Green map:

$$
y = G_1 v \iff \{\Delta^2 y = 0 \text{ in } \Omega; \ y|_\Gamma = v, \ \frac{\partial y}{\partial \nu}\Big|_\Gamma = 0\}, \qquad (7.28)
$$

and the spaces

$$
Y = H^{-1}(\Omega) \times V' = [\mathscr{D}(A^{\frac{1}{4}})]' \times [\mathscr{D}(A^{\frac{3}{4}})]'. \qquad (7.29)
$$

where with equivalent norms

$$
\mathscr{D}(A^{\frac{1}{4}}) = H^1_0(\Omega); \quad \mathscr{D}(A^{\frac{3}{4}}) = V. \qquad (7.30)
$$

__Assumption (1.3)__: $(-A)^{-\gamma} B \in \mathscr{L}(U,Y)$. It is plainly satisfied with, say, $\gamma = 1$, as one sees by proceeding as in (7.19).

Assumption (H.2) = (1.6). One can show that [L-T.9], [F-L-T, Appendix C], with y = [y_1, y_2],

$$B^* e^{A^* t} \left| \begin{matrix} y_1 \\ y_2 \end{matrix} \right| = \frac{\partial \Delta \, \phi(t)}{\partial \nu}, \quad y \in Y, \tag{7.31}$$

where $\phi(t) = \phi(t, \phi_0, \phi_1)$ solves the corresponding homogeneous problem (7.21), this time however with initial data,

$$\phi_0 = A^{-\frac{1}{2}} y_2 \in \mathcal{D}(A^{\frac{1}{2}}) = V; \quad \phi_1 = -A^{-\frac{1}{2}} y_1 \in \mathcal{D}(A^{\frac{1}{2}}) = H_0^1(\Omega). \tag{7.32}$$

Thus, by (7.31) and (7.32), an equivalent formulation of assumption (H.2) = (1.6) is the inequality,

$$\int_\Sigma \left(\frac{\partial \Delta \phi}{\partial \nu} \right)^2 d\Sigma \leq c_T \| \{\phi_0, \phi_1\} \|^2_{V \times H_0^1(\Omega)}, \tag{7.33}$$

for the trace of the solution to problem (7.21), (7.32). Again, as in prior cases, inequality (7.33) does NOT follow from *a priori* (optimal) interior regularity $\phi(t) \in C([0,T];V)$ of the solution to problem (7.21), (7.32). It is an independent regularity result which holds indeed true [Lio.2], [L-T.11], for any general smooth Ω. Thus assumption (H.2) = (1.6) holds true for problem (7.24).

Finite Cost Condition (1.9). The same considerations apply now as in the case of the wave equation in Section 7.1 and of the Euler-Bernoulli problem (7.13). Exact controllability of problem (7.24) on the state space $H^{-1}(\Omega) \times V'$ holds true for any $T > 0$ within the class of $L_2(\Sigma)$-controls u, at least under some geometrical conditions on Ω. To eliminate geometrical conditions, one may add, however, a suitable second control in the B.C. (7.24d). This and more is proved in [L-T.11]. Uniform stabilization results on $H^{-1}(\Omega) \times V'$ can also be given under the same conditions as exact controllability results [B-T], [T.7]: i.e., under some geometrical conditions on Ω if only one feedback in the B.C. (7.24c) is used; or else with no geometrical conditions on Ω if two feedbacks are used in the B.C. (7.24c) and (7.24d) respectively. In any case, the Finite Cost Condition (1.9) holds true for problem (7.24) under some geometrical conditions on Ω, or else with no geometrical conditions on Ω, if one adds a second control in (7.24d) and in the cost (7.25).

<u>Detectability Conditions (5.17)-(5.19)</u>. This holds true since R = I.

<u>Conclusion</u>: T = ∞.   Theorem 5.2 applies also to problem (7.24).

<u>Conclusion</u>: T < ∞.   Theorem 3.1 applies to (7.24) with R = I while Theorem 3.3 require additional regularity on R (Remark 3.2).

<u>Case 3</u>.   We now consider on any smooth $\Omega \subset R^n$,

$$
\begin{cases}
w_{tt} + \Delta^2 w = 0 & \text{in } Q; & (7.34a) \\
w(0,\cdot) = w_0, \ w_t(0,\cdot) = w_1 & \text{in } \Omega; & (7.34b) \\
w|_\Sigma = 0 & \text{in } \Sigma; & (7.34c) \\
\Delta w|_\Sigma = u & \text{in } \Sigma, & (7.34d)
\end{cases}
$$

with boundary control $u \in L_2(\Sigma)$.   Consistently with optimal regularity theory [L-T.14], [Lio.2], we take the following cost functional

$$
J(u,w) = \int_0^\infty \{ \|w(t)\|^2_{H^1(\Omega)} + \|w_t(t)\|^2_{H^{-1}(\Omega)} + |u(t)|^2_{L_2(\Gamma)} \} dt \quad (7.35)
$$

with initial data $\{w_0, w_1\} \in H_0^1(\Omega) \times H^{-1}(\Omega)$.

<u>Abstract setting</u>.   To put problem (7.34), (7.35) into the abstract model (1.1), (1.2), we introduce the operators

$$
Ah = \Delta^2 h, \ \mathcal{D}(A) = \{ h \in H^4(\Omega): h|_\Gamma = \Delta h|_\Gamma = 0 \}; \quad (7.36a)
$$

$$
A = \begin{vmatrix} 0 & I \\ -A & 0 \end{vmatrix}; \quad Bu = \begin{vmatrix} 0 \\ AG_4 u \end{vmatrix}, \quad R = I, \quad (7.36b)
$$

where G_4 is the appropriate Green map

$$
y = G_4 v \iff \{ \Delta^2 y = 0 \text{ in } \Omega; \ y|_\Gamma = 0, \ \Delta y|_\Gamma = v \}, \quad (7.37)
$$

and the spaces

$$
Y = H_0^1(\Omega) \times H^{-1}(\Omega) = \mathcal{D}(A^{\frac{1}{4}}) \times [\mathcal{D}(A^{\frac{1}{4}})]'; \quad U = L_2(\Gamma). \quad (7.38)
$$

<u>Assumption (1.3)</u>: $(-A)^{-\gamma} B \in \mathcal{L}(U,Y)$.   It is plainly satisfied with, say, $\gamma = 1$, as one sees by proceeding as in (7.19), since $G_4 \in \mathcal{L}(L_2(\Gamma), L_2(\Omega))$.

Assumption (H.2) = (1.6). One can show that [L-T.14], [L-T.15],

$$B^* \begin{vmatrix} y_1 \\ y_2 \end{vmatrix} = G_4^* A^{\frac{1}{2}} y_2; \qquad B^* e^{A^* t} \begin{vmatrix} y_1 \\ y_2 \end{vmatrix} = \frac{\partial(\Delta\phi(t))}{\partial\nu}, \qquad (7.39)$$

where $\phi(t) = \phi(t, \phi_0, \phi_1)$ solves the corresponding homogeneous problem,

$$\begin{cases} \phi_{tt} + \Delta^2 \phi = 0 & \text{in } Q; & (7.40a) \\ \phi(0, \cdot) = \phi_0, \ \phi_t(0, \cdot) = \phi_1 & \text{in } \Omega; & (7.40b) \\ \phi|_{\Sigma} = \Delta\phi|_{\Sigma} \equiv 0 & \text{in } \Sigma, & (7.40c) \end{cases}$$

with

$$\phi_0 = A^{-1} y_2 \in \mathcal{D}(A^{\frac{3}{4}}); \quad \phi_1 = y_1 \in \mathcal{D}(A^{\frac{1}{4}}). \qquad (7.41)$$

Thus, by (7.39), (7.41), an equivalent formulation of assumption (H.2) = (1.6) is the inequality

$$\int_{\Sigma} \left(\frac{\partial(\Delta\phi)}{\partial\nu} \right)^2 d\Sigma \leq c_T \| \{\phi_0, \phi_1\} \|^2_{\mathcal{D}(A^{\frac{3}{4}}) \times \mathcal{D}(A^{\frac{1}{4}})} \qquad (7.42)$$

for the trace of the solution of problem (7.40). As in preceding cases, it should be noted that inequality (7.42) does NOT follow from *a priori* (optimal) interior regularity $\phi(t) \in C([0,T]; \mathcal{D}(A^{\frac{3}{4}}))$, where the equivalent norms [Gr.1]

$$\mathcal{D}(A^{\frac{3}{4}}) = V = \{h \in H^3(\Omega): h|_{\Gamma} = \Delta h|_{\Gamma} = 0\}. \qquad (7.43)$$

Inequality (7.42) is an independent regularity result which holds indeed true [L-T.14], [Lio.2] for any general smooth Ω. Thus, assumption (H.2) = (1.6) holds true for problem (7.40).

Finite Cost Condition (1.9). Originally, exact controllability on any [0,T] for the Euler-Bernoulli equation (7.34a) was shown on the required space (7.38) by using however two controls: $w|_{\Sigma} = u_1 \in H_0^1(0,T;L_2(\Gamma))$ and $\Delta w|_{\Sigma} = u_2 \in L_2(\Sigma)$ [L-T.15; Thm. 1.2], [Lio.2]. This is equivalent to the inequality

$$\int_{\Sigma} \left(\frac{\partial(\Delta\phi)}{\partial\nu} \right)^2 + \left(\frac{\partial\phi_t}{\partial\nu} \right)^2 d\Sigma \geq c_T \| \{\phi_0, \phi_1\} \|^2_{\mathcal{D}(A^{\frac{3}{4}}) \times \mathcal{D}(A^{\frac{1}{4}})} \qquad (7.44)$$

for problem (7.40), (7.41), see [L-T.15, Lemma 3.2]. It was later observed in [Leb.1] that the term $\frac{\partial \phi_t}{\partial \nu} \in L_2(\Sigma)$ in (7.44) can be 'absorbed.' (This is a non-trivial improvement, since $\frac{\partial \phi_t}{\partial \nu}$ in $L_2(\Sigma)$ is not a lower order term with respect to $\frac{\partial(\Delta\phi)}{\partial \nu}$ in $L_2(\Sigma)$, and hence the usual compactness/uniqueness argument does not apply). This improvement has the important equivalent formulation that problem (7.34) with just one control $u \in L_2(\Sigma)$ in (7.34d) is indeed exactly controllable on [0,T] on the space (7.38) as desired. Thus, the required Finite Cost Condition (1.9) does hold true for problem (7.34). Indeed, the idea in [Leb.1] is useful also in the proof which establishes the corresponding uniform stabilization result for problem (7.34) [Las.7]. Exact controllability has also been extended to the two-dimensional plate model with physical moment [H.1].

Detectability Condition (5.17)-(5.19). This holds true since R = I.

Conclusion: T = ∞. Theorem 5.2 applies to problem (7.34), (7.35).

Conclusion: T < ∞. Same as in Case 2.

Case 4. We now consider on any smooth $\Omega \in R^n$,

$$
\begin{cases}
w_{tt} + \Delta^2 w = 0 & \text{in } Q; & (7.45a) \\
w(0,\cdot) = w_0; \; w_t(0,\cdot) = w_1 & \text{in } \Omega; & (7.45b) \\
w|_\Sigma = u & \text{in } \Sigma; & (7.45c) \\
\Delta w|_\Sigma = 0 & \text{in } \Sigma, & (7.45d)
\end{cases}
$$

with boundary control $u \in L_2(\Sigma)$. Consistently with optimal regularity theory [L-T.15], [Lio.2], the cost functional to be minimized for $\{w_0, w_1\} \in H^{-1}(\Omega) \times V'$ is:

$$
J(u,w) = \int_0^\infty \{ \|w(t)\|^2_{H^{-1}(\Omega)} + \|w_t(t)\|^2_{V'} + \|u(t)\|^2_{L_2(\Gamma)} \} dt, \quad (7.46)
$$

where V' is the dual of V in (7.43), and where we note that $\mathcal{D}(A^{\frac{1}{4}}) = H_0^1(\Omega)$, hence (with equivalent norms)

$$
Y = H^{-1}(\Omega) \times V' = [\mathcal{D}(A^{\frac{1}{4}})]' \times [\mathcal{D}(A^{\frac{1}{4}})]' \quad (7.47)
$$

with A the operator defined in (7.36a). By making the change of variable $\varsigma = -A^{\frac{1}{2}}w$, w solution of problem (7.45), we find that the new variable ς satisfies precisely problem (7.34), and that the space in (7.47) is mapped into the space in (7.38). Thus, with $U = L_2(\Gamma)$, problem (7.45), (7.46) for w is converted into a problem for ς which satisfies (7.34), (7.35). Thus, all the required properties (regularity, exact controllability, uniform stabilization) for (7.45), (7.46), are equivalent to the corresponding properties for (7.34), (7.35). Hence, Theorem 5.2 applies also to problem (7.45), (7.46).

For sake of completeness we note now that A is the same as in (7.36b), while B is now

$$Bu = \begin{vmatrix} 0 \\ AG_3 u \end{vmatrix}; \quad y = G_3 v \iff \{\Delta y^2 = 0 \text{ in } \Omega; \; y|_\Gamma = v; \; \Delta y|_\Gamma = 0\}; \quad (7.48)$$

$$B^* \begin{vmatrix} y_1 \\ y_2 \end{vmatrix} = G_3^* A^{-\frac{1}{2}} y_2; \quad B^* e^{A^* t} \begin{vmatrix} y_1 \\ y_2 \end{vmatrix} = \frac{\partial(\Delta \phi(t))}{\partial \nu}, \quad (7.49)$$

where $\phi(t)$ is the solution of the same problem (7.40) with

$$\phi_0 = A^{-\frac{3}{2}} y_2 \in \mathcal{D}(A^{\frac{1}{4}}); \quad \phi_1 = A^{-\frac{1}{2}} y_1 \in \mathcal{D}(A^{\frac{1}{4}}), \quad (7.50)$$

i.e., the same regularity of the initial data as in (7.41).

7.3.   Class (H.2): Schrödinger equation with Dirichlet boundary control

In $\Omega \subset R^n$ we consider the Schrödinger equation

$$\begin{cases} y_t = -i\Delta y & \text{in } Q; & (7.51a) \\ y(0, \cdot) = y_0 & \text{in } \Omega; & (7.51b) \\ y|_\Sigma = u & \text{in } \Sigma, & (7.51c) \end{cases}$$

with control $u \in L_2(\Sigma)$. Consistently with the optimal regularity theory proved in [L-T.25], we take $y_0 \in H^{-1}(\Omega)$ and the cost functional

$$J(u, y) = \int_0^\infty \{\|y(t)\|^2_{H^{-1}(\Omega)} + \|u(t)\|^2_{L_2(\Gamma)}\} dt. \quad (7.52)$$

Abstract setting. [L-T.25] To put problem (7.51), (7.52) into the abstract model (1.1), (1.2), we take the following operators and spaces where D is the same Dirichlet map introduced in Eq. (3.47):

$$A = iA, \quad A = -\Delta; \quad \mathcal{D}(A) = H^2(\Omega) \cap H^1_0(\Omega); \qquad (7.53)$$

$$B = -iAD; \quad R = I; \quad B^* e^{A^* t} y = -i \frac{\partial \phi(t)}{\partial \nu}; \qquad (7.54)$$

where $\phi(t) = \phi(t, \phi 0)$ solves the following homogeneous problem:

$$\begin{cases} \phi_t = i\Delta\phi & \text{in } Q; & (7.55a) \\ \phi(0, \cdot) = \phi_0 & \text{in } \Omega; & (7.55b) \\ \phi|_\Sigma = 0 & \text{in } \Sigma; & (7.55c) \end{cases}$$

$$Y = H^{-1}(\Omega); \quad U = L_2(\Gamma). \qquad (7.56)$$

Assumption (1.3): $(-A)^{-\gamma} B \in \mathcal{L}(U, Y)$. This is plainly true, say with $\gamma = 1$, by (7.53), (7.54) via property (6.48) for D.

Assumption (H.2) = (1.6). According to (7.54), an equivalent formulation for assumption (H.2) = (1.6) is the inequality

$$\int_\Sigma \left| \frac{\partial \phi}{\partial \nu} \right|^2 d\Sigma \leq c_T \|\phi_0\|^2_{H^1_0(\Omega)} \qquad (7.57)$$

for the trace of the solution of (7.55). As in preceding cases, we note that inequality (7.57) does NOT follow from *a-priori* optimal interior regularity $\phi(t) \in C([0,T]; H^1_0(\Omega))$ with $\phi_0 \in H^1_0(\Omega)$. Inequality (7.57) is an independent regularity result, which is established in [L-T.25].

Finite Cost Condition (1.9). Both exact controllability and uniform stabilization of problem (7.51) on the space $H^{-1}(\Omega)$ on any $[0,T]$, with $L_2(0,T; L_2(\Gamma))$-controls as well as uniform stabilization on the same space $H^{-1}(\Omega)$ with $L_2(0,\infty; L_1(\Gamma))$ feedback controls hold true without geometrical conditions on Ω, if the control action is exercised on all of Γ. See [Leb.1], [L-T.25] for exact controllability and [L-T.25] for uniform stabilization. Thus, *a fortiori* the Finite Cost Condition (1.9) holds true for problem (7.51), (7.52).

Detectability Conditions (5.17)-(5.19). This holds true since R = I.

<u>Conclusion</u>: T = ∞.   Theorem 5.2 applies to problem (7.51), (7.52).

<u>Conclusion</u>: T < ∞.   Theorem 3.1 applies to problem (7.51), (7.52),
where R = I.   Theorem 3.3 requires additional smoothing on R.

7.4.   <u>Class (H.2): First-order hyperbolic systems</u>

Consider the following not necessarily symmetric or dissipative
first order hyperbolic system in the unknown $y(\xi_1, \xi_2, \ldots, \xi_n) \in R^m$

$$
\begin{cases}
\partial_t y = \sum_{j=0}^{n} A_j(\xi) \partial_j y & \text{in } (0,T] \times \Omega; & (7.58a) \\
y|_{t=0} = y_0 \in [L_2(\Omega)]^m & \text{in } \Omega; & (7.58b) \\
M(\sigma) y(t,\sigma) = u(t,\sigma) \in L_2(0,T;[L_2(\Gamma)]^k) & \text{in } (0,T) \times \Gamma, & (7.58c)
\end{cases}
$$

where A_j, respect. M, are smooth m×m, respect. k×m, matrix valued
functions under the assumptions of (a) strict hyperbolicity, and of (b)
Γ being non-characteristic, and (c) rank $M(\sigma) = k \leq m$; here k stands
for the number of negative eigenvalues of $A_N = \sum_{j=1}^{n} A_j N_j$, $N = [N_1, \ldots, N_m]$
outward unit normal.   Without loss of generality, we may assume that
(after a similarity transformation)

$$
\begin{aligned}
M = [I,S] \quad & I: \text{k×k identity}; \quad A_N = \text{diag}[A_N^-, A_N^+], \quad A_N^- < 0, \quad A_N^+ > 0, \\
& S: \text{k×(m-k)}
\end{aligned}
$$
(7.59)

where A_N^- is a k×k matrix having the same negative eigenvalues of A_N;
and A_N^+ is a (m-k)×(m-k) matrix having the same positive eigenvalues of
A_N.   With (7.58) we associate the cost

$$
J(u,y) = \int_0^T \left\{ \|Ry(t)\|^2_{[L_2(\Omega)]^m} + |u(t)|_{[L_2(\Gamma)]^k} \right\} dt \qquad (7.60)
$$

for T < ∞ or T = ∞, with $R \in \mathcal{L}([L_2(\Omega)]^m)$.

<u>Abstract setting</u>.   [C-L], [D-L-S]   To put problem (7.58) in the
abstract form (1.1), (1.2), we choose

$$
Z = Y = [L_2(\Omega)]^m; \qquad U = [L_2(\Gamma)]^k; \qquad (7.61)
$$

A = first order differential operator F with
homogeneous boundary conditions, (7.62)

where

$$Fy = \sum_{j=0}^{n} A_j(\xi)\partial_t y; \quad B = AD_1; \quad A^{-1}B = D_1;$$ (7.63)

where (up to a translation)

$$D_1 g = f \iff \{Ff = 0 \text{ in } \Omega; \quad Mf = g \text{ in } \Gamma\};$$ (7.64)

$$D_1: \text{continuous } [L_2(\Gamma)]^k \to [L_2(\Omega)]^m.$$ (7.65)

It is well known that A generates a s.c. semigroup e^{At} on Y.

Assumption (1.3): $(-A)^{-\gamma}B \in \mathcal{L}(U;Y)$. This follows with $\gamma = 1$ from (7.63)-(7.64).

Assumption (H.2) = (1.6). The required regularity property [H.2] is available from [Kr.1], [Rau.1], and is put in a semigroup framework in [C-L]. These formulas will be needed in the numerical approximation treatments of Section 10.4 in Part II. We have [C-L],

$$(Lu)(t) = A\int_0^t \exp[A(t-\tau)]D_1 u(\tau)d\tau;$$ (7.66)

$$B^*x = A_N^- x^-|_\Gamma, \quad x = [x^-, x^+], \quad \dim x^- = k; \quad \dim x^+ = m-k.$$ (7.67)

It is readily verified that the Y-adjoint A^* of A is

$$A^*y = -\sum_{j=1}^{n} A_j^T(\xi)\partial_j y - \sum_{j=0}^{n} \partial_j A_j^T(\xi)y + A_0^T(\xi);$$ (7.68)

$$\mathcal{D}(A^*) = \{h \in [L^2(\Omega)]^m; \ A^*h \in [L_2(\Omega)]^m \text{ and } M^*h = 0\};$$ (7.69)

$$M^* = [-(A_N^T)^{-1}S^T A_N^-, I_{m-k}].$$ (7.70)

The regularity results of [Kr.1] for strictly hyperbolic non-characteristic systems yield

$$\|e^{A^*t}x\|_{L_2(0,T;[L_2(\Omega)]^m)} \leq c_T\|x\|_{[L_2(\Omega)]^m},$$ (7.71)

which combined with (7.67) yields the desired estimate (H.2) as in [Rau.1]

$$\|B^* e^{A^* t} x\|_{L_2(0,T;[L_2(\Gamma)]^k)} \leq c_T \|x\|_{[L_2(\Omega)]^m} . \qquad (7.72)$$

Finite Cost Condition (1.9). A sufficient condition for the Finite Cost Condition (1.9) to hold is the following 'exact null controllability' property: there exists T > 0 such that for any $Y_0 \in [L_2(\Omega)]^m$, there exists $u \in L_2(0,T;[L_2(\Gamma)]^k)$ such that the corresponding solution of problem (7.58) satisfies Y(T) = 0. This property has been proved in the one-dimensional case where then Ω is a finite open interval [Ru.2, Thm. 3.2].

Detectability Condition (5.17)-(5.19): holds true with, say, R = I.

Conclusion: T = ∞. Theorem 5.2 on the ARE applies to problem (8.58) with R = I and Ω one dimensional. Progress on null controllability is needed to apply Theorem 5.2 also in the multidimensional case. ∎

Conclusion: T < ∞. Theorem 3.1 applies to problem (7.58), (7.60) for any $R \in \mathcal{L}([L_2(\Omega)]^m)$ and yields a "viscosity solution" (Section 3.4) Riccati operator P(t) from (3.6). That this P(t) solves the DRE (3.21) for all $x, z \in \mathcal{D}(A)$ requires one additional minimal assumption; e.g., that $R^* R A^\varepsilon \in \mathcal{L}(Y)$ [C-L]. Finally, uniqueness as in Theorem 3.3 requires additional smoothing; e.g., $R^* R A \in \mathcal{L}(Y)$ [D-L-T, p. 34] to satisfy assumption (3.25). Below, we shall verify directly (3.25) in two cases, where then the existence and uniqueness Theorem 3.3 applies and yields, in particular, that

$$B^* P(t) x = A_N^- [P(t) x]^- |_\Gamma : \text{continuous } [L_2(\Gamma)]^k \to C([0,T];[L_2(\Omega)]^m) .$$

$$(7.73)$$

Case 1. Take R to be a bounded, 1-rank operator on Y: Rx = (x,a)b, $a, b \in Y$, so that $R^* x = (x,b)a$ and $R^* R x = (x,c)c$, c = ∥b∥a. Then, for $u \in U$, $R^* R e^{At} B u = (u, B^* e^{A^* t} c) c$ and

$$\int_0^T \|R^* R e^{At} B u\|_Y dt \leq \|c\| \|u\|_U \sqrt{T} \|B^* e^{A^* t} c\|_{L_2(0,T;U)}$$

$$\leq \text{const}_T \|c\| \|u\|_U , \qquad (7.74)$$

and (3.25) holds true. Thus Theorem 3.3 applies. Note that we have taken any a,b ∈ Y, which would not satisfy the sufficient condition $R^*RA \in \mathcal{L}(Y)$ in general.

Case 2. We now take

$$R^*R: \text{ continuous } [L_2(\Omega)]^m \to [H_0^{1/2+\varepsilon}(\Omega)]^m. \qquad (7.75)$$

The differentiability theorem in [Rau.1] gives

$$e^{A^*t}: \text{ continuous } [H_0^{1/2+\varepsilon}(\Omega)]^m \to C([0,T];[H^{1/2+\varepsilon}(\Omega)]^m). \qquad (7.76)$$

Instead of proving ((3.25): $R^*Re^{At}B \in \mathcal{L}(U;L_1(0,T;Y))$, we shall equivalently prove the dual statement, see (7.61),

$$g \to \int_0^T B^*e^{A^*t}R^*Rg(t)dt: \text{ continuous } L_\infty(0,T;Y) \to U. \qquad (7.77)$$

Indeed, for such g we use (7.67) with A_N^- invertible and standard trace theory to obtain

$$\left\| \int_0^T B^*e^{A^*t}R^*Rg(t)dt \right\|_U \leq c_T \sup_{0 \leq t \leq T} \| e^{A^*t}R^*Rg(t) \|_{[H^{1/2+\varepsilon}(\Omega)]^m}$$

(by (7.26))

$$\leq c_T \sup_{0 \leq t \leq T} \| R^*Rg(t) \|_{[H_0^{1/2+\varepsilon}(\Omega)]^m}$$

(by (7.75))

$$\leq c_T \sup \| g(t) \|_{[L_2(\Omega)]^m} = c_T \| g \|_{L_\infty(0,T;Y)}$$

as desired, and (7.77) is proved.

7.5. Class (H.2): Kirchoff plate with boundary control in the bending moment

In $\Omega \subset R^n$, we consider the Kirchoff plate

$$\begin{cases} w_{tt} - \rho\Delta w_{tt} + \Delta^2 w = 0 & \text{in } (0,T]\times\Omega \equiv Q, \qquad (7.78a) \\ w(0,\cdot) = w_0, \ w_t(0,\cdot) = w_1 & \text{in } \Omega, \qquad (7.78b) \\ w|_\Sigma \equiv 0 & \text{in } (0,T]\times\Gamma, \qquad (7.78c) \\ \Delta w|_\Sigma = u & \text{in } \Sigma, \qquad (7.78d) \end{cases}$$

with $\rho > 0$ a constant, and with just one boundary control u which we take in $L_2(\Sigma)$. Optimal regularity theory of problem (7.78) is given in [L-T.16]. Consistently with these results, we take the following cost functional

$$J(u,w) = \int_0^\infty \|w(t)\|^2_{H^2(\Omega)} + \|w_t(t)\|^2_{H^1(\Omega)} + \|u(t)\|^2_{L_2(\Gamma)} dt \qquad (7.79)$$

with initial data $\{w_0, w_1\} \in [H^2(\Omega) \cap H^1_0(\Omega)] \times H^1_0(\Omega)$.

Abstract setting. To put problem (7.78), (7.79) into the abstract model (1.1), (1.2), we introduce the positive self-adjoint operators

$$Ah = \Delta^2 h, \quad \mathcal{D}(A) = \{h \in H^4(\Omega); \ h|_\Gamma = \Delta h|_\Gamma = 0\}$$

$$A^{\frac{1}{2}} h = -\Delta h, \quad \mathcal{D}(A^{\frac{1}{2}}) = H^2(\Omega) \cap H^1_0(\Omega)$$

the same as in (6.27) of Example 6.1, and define the operators

$$\mathcal{A} = \begin{vmatrix} 0 & I \\ -A & 0 \end{vmatrix}; \quad Bu = \begin{vmatrix} 0 \\ AG_2 u \end{vmatrix}; \quad A = (I + \rho A^{\frac{1}{2}})^{-1} A; \quad R = I, \qquad (7.80)$$

and G_2 is the same Green map defined in (6.45) in Example 6.3, which satisfies Eq. (6.47) in terms of the Dirichlet map D: $G_2 = -A^{-\frac{1}{2}} D$, with D as in (6.48). We also define the spaces

$$Y = [H^2(\Omega) \cap H^1_0(\Omega)] \times H^1_0(\Omega) = \mathcal{D}(A^{\frac{1}{2}}) \times \mathcal{D}(A^{\frac{1}{4}}); \quad U = L_2(\Gamma). \qquad (7.81)$$

Assumption (1.3): $(-\mathcal{A})^{-\gamma} B \in \mathcal{L}(U,Y)$. By (7.80) and (7.81) with $u \in L_2(\Gamma)$, we plainly have

$$(-\mathcal{A})^{-1} Bu = \begin{vmatrix} 0 & A^{-1} \\ -I & 0 \end{vmatrix} \begin{vmatrix} 0 \\ AG_2 u \end{vmatrix} = \begin{vmatrix} G_2 u \\ 0 \end{vmatrix} = \begin{vmatrix} -A^{-\frac{1}{2}} Du \\ 0 \end{vmatrix} \in Y, \qquad (7.82)$$

and assumption (1.3) holds true for problem (7.78).

Assumption (H.2) = (1.6). One can show that [L-T.16]

$$B^* e^{A^* t} \begin{vmatrix} z_1 \\ z_2 \end{vmatrix} = \frac{\partial \Delta \phi(t)}{\partial \nu} \Big|_\Sigma , \qquad (7.83)$$

where $\phi(t) = \phi(t, \phi_0, \phi_1)$ solves the corresponding homogeneous problem

$$\begin{cases} \phi_{tt} - \rho \Delta \phi_{tt} + \Delta^2 \phi = 0 & (7.84a) \\ \phi(0, \cdot) = \phi_0, \ \phi_t(0, \cdot) = \phi_1 & (7.84b) \\ \phi|_\Sigma \equiv \Delta \phi|_\Sigma \equiv 0 & (7.84c) \end{cases}$$

with

$$\phi_0 = (I + \rho A^{\frac{1}{2}})^{-1} z_2 \in \mathcal{D}(A^{\frac{3}{4}}); \qquad (7.85a)$$

$$\phi_1 = -(I + \rho A^{\frac{1}{2}})^{-1} A^{\frac{1}{2}} z_1 \in \mathcal{D}(A^{\frac{1}{2}}). \qquad (7.85b)$$

Thus, by (7.83), (7.85), an equivalent formulation of assumption (H.2) = (1.6) is the inequality

$$\int_\Sigma \left(\frac{\partial \Delta \phi}{\partial \nu} \right)^2 d\Sigma \leq c_T \| \{\phi_0, \phi_1\} \|^2_{\mathcal{D}(A^{\frac{3}{4}}) \times \mathcal{D}(A^{\frac{1}{2}})}. \qquad (7.86)$$

This inequality holds indeed true, as recently shown in [L-T.16] by multiplier methods. Thus, assumption (H.2) = (1.6) holds true for problem (7.78) for general smooth Ω.

Finite Cost Condition (1.9). The same considerations as in Section 7.1 for the wave equation and Section 7.2 for the Euler-Bernoulli equation apply. It was recently proved that problem (7.78) is exactly controllable for sufficiently large $T > 0$ on the state space $Y = [H^2(\Omega) \cap H^1_0(\Omega)] \times H^1_0(\Omega)$ within the class of $L_2(\Sigma)$-controls u, with no geometrical conditions on Ω (except Γ smooth), if u is applied to all of Γ [L-T.16]. As a consequence, the F.C.C. (1.9) holds true. (Problem (7.78) is also uniformly stabilizable under some geometrical conditions on Ω, e.g., strict convexity [L-T.16].)

Detectability Condition (5.17)-(5.19). This holds true since $R = I$.

Conclusion: $T = \infty$. Theorem 5.2 applies to problems (7.78), (7.79).

Conclusion: $T < \infty$. Theorem 3.1 applies to problem (7.78), (7.79), where $R = I$, while Theorem 3.3 requires additional smoothing on R.

7.6. Class (H.2): A two-dimensional plate model with boundary control as a bending moment

We return to the Euler-Bernoulli equation of Section 7.2, except that now $\Omega \subset R^2$ is a smooth two-dimensional domain and now the control u acts as a (physical) bending moment. More precisely, we consider the problem

$$
\begin{cases}
w_{tt} + \Delta^2 w = 0 & \text{in } (0,T] \times \Omega = Q; & (7.87a) \\
w(0,\cdot) = w_0, \; w_t(0,\cdot) = w_1 & \text{in } \Omega; & (7.87b) \\
w|_\Sigma = 0 & \text{in } (0,T] \times \Gamma = \Sigma; & (7.87c) \\
[\Delta w + (1-\mu)B_1 w]_\Sigma = u & \text{in } \Sigma, & (7.87d)
\end{cases}
$$

with boundary control $u \in L_2(\Sigma)$. In (7.87d) the constant μ is $0 \leq \mu < 1$ (physically $0 < \mu < \frac{1}{2}$) while the boundary operator B_1 takes the form

$$
B_1 w = - \frac{\partial^2 w}{\partial \tau^2} - k \frac{\partial w}{\partial \nu} = - k \frac{\partial w}{\partial \nu} , \qquad (7.88)
$$

$k(x)$ being the curvature, as the tangential derivative vanishes by (7.87c). Consistently with optimal regularity theory as in [L-T.14], we take the following cost functional

$$
J(u,w) = \int_0^\infty \{ \|w(t)\|^2_{H^1(\Omega)} + \|w_t(t)\|^2_{H^{-1}(\Omega)} + |u(t)|^2_{L_2(\Gamma_1)} \} dt \qquad (7.89)
$$

with initial data $\{w_0, w_1\} \in H^1_0(\Omega) \times H^{-1}(\Omega)$.

Abstract setting. To put problem (7.87)-(7.89) into the abstract model (1.1), (1.2), we introduce the positive, self-adjoint operator

$$
Ah \equiv \Delta^2 h, \quad \mathcal{D}(A) = \{ u \in H^4(\Omega): h|_\Gamma = 0, \; \Delta h + (1-\mu)B_1 h|_\Gamma = 0 \}, \qquad (7.90)
$$

and define the operators

$$
A = \begin{bmatrix} 0 & I \\ -A & 0 \end{bmatrix}; \quad Bu = \begin{bmatrix} 0 \\ AGu \end{bmatrix}; \quad R = I, \qquad (7.91)
$$

where G is the appropriate Green map defined by

$$
y = Gv \iff \{ \Delta^2 y = 0 \text{ in } \Omega; \; y|_\Gamma = 0; \; [\Delta y + (1-\mu)B_1 y]_\Gamma = v \}, \qquad (7.92)
$$

and the spaces

$$Y = H_0^1(\Omega)H^{-1}(\Omega) = \mathcal{D}(A^{\frac{1}{4}})\times[\mathcal{D}(A^{\frac{1}{4}})]'; \quad U = L_2(\Gamma). \qquad (7.93)$$

Assumption (1.3). $(-A)^{-\gamma}B \in L(U,Y)$. Since G is certainly bounded $L_2(\Gamma) \to L_2(\Omega)$, we readily obtain from (7.91) with $u \in L_2(\Gamma)$:

$$(-A)^{-1}Bu = \begin{bmatrix} 0 & A^{-1} \\ -I & 0 \end{bmatrix} \begin{bmatrix} 0 \\ AGu \end{bmatrix} = \begin{bmatrix} Gu \\ 0 \end{bmatrix} \in Y, \qquad (7.94)$$

and Assumption (1.3) holds true for problem (7.87) with $\gamma = 1$.

Assumption (H.2) = (1.6). One can show [Hor.1] that as in (7.39) for problem (7.34) we have

$$B^*e^{A^*t}\begin{bmatrix} y_1 \\ y_2 \end{bmatrix} = -\frac{\partial}{\partial\nu}\Delta\phi(t)|_\Sigma, \quad y = [y_1,y_2] \in Y, \qquad (7.95)$$

where $\phi(t) = \phi(t,\phi_0,\phi_1)$ solves the corresponding problem with nonhomogeneous boundary conditions

$$\begin{cases} \phi_{tt}+\Delta^2\phi = (1-\mu)D(k\frac{\partial}{\partial\nu}(\Delta\phi)) & \text{in } Q; & (7.96a) \\ \phi(0,\cdot) = \phi_0 \in \mathcal{D}(A_D^{\frac{3}{4}}) \\ \phi_t(0,\cdot) = \phi_1 \in \mathcal{D}(A_D^{\frac{1}{4}}) \end{cases} & \text{in } \Omega; & (7.96b) \\ \phi = 0 & \text{on } \Sigma; & (7.96c) \\ \Delta\phi = 0 & \text{on } \Sigma, & (7.96d) \end{cases}$$

with

$$\phi_0 = A_D^{-1}A^{-\frac{1}{4}}y_2 \in \mathcal{D}(A_D^{\frac{3}{4}}) = H^3(\Omega)\cap H_0^1(\Omega);$$

$$\phi_1 = A_D^{-1}A^{\frac{1}{4}}y_1 \in \mathcal{D}(A_D^{\frac{1}{4}}) = H^{-1}(\Omega), \qquad (7.97)$$

where A_D is the positive self-adjoint operator defined by

$$A_Dh = -\Delta h, \quad \mathcal{D}(A_D) = H^2(\Omega)\cap H_0^1(\Omega), \qquad (7.98)$$

and D is the Dirichlet map as defined in (7.4). Thus, by (7.95), (7.96), an equivalent formulation of Assumption (H.2) = (1.6) is the inequality

$$\int_{\Sigma}\left(\frac{\partial}{\partial\nu}\,\Delta\phi\right)^2 d\Gamma\ dt \le C_T\|\{\phi_0,\phi_1\}\|^2_{[H^3(\Omega)\cap H^1_0(\Omega)]\times H^1_0(\Omega)} \qquad (7.99)$$

for the trace of the solution to problem (7.96). As in previous cases, it should be noted that inequality (7.99) does NOT follow from *a priori* (optimal) interior regularity of the solution to the problem (7.96), (7.97). It is an independent regularity result which holds true [Hor.1], for any general smooth Ω. Thus Assumption (H.2) = (1.6) holds true for problem (7.87).

Finite Cost Condition. Exact controllability on any [0,T] for the Euler-Bernoulli plate (7.87) on the state space $Y = H^1_0(\Omega)\times H^{-1}(\Omega)$ within the class of $L_2(\Sigma)$-controls u is equivalent to the inequality

$$\int_{\Sigma}\left(\frac{\partial(\Delta\phi)}{\partial\nu}\right)^2 d\Sigma \ge C_T\|\{\phi_0,\phi_1\}\|^2_{[H^3(\Omega)\cap H^1_0(\Omega)]\times H^1_0(\Omega)}, \qquad (7.100)$$

which indeed holds true with no geometrical conditions on Ω, if the control u acts, as assumed, on all of Γ [Hor.1] (in this reference, an extension of this result which allows the control functions to act only on a portion of the boundary may also be found: in this case, of course, geometrical conditions are needed). As in the case of the Euler-Bernoulli problem (7.34) (Case 3), a novel difficulty is to absorb the term $\frac{\partial}{\partial\nu}\,\phi_t \in L_2(\Sigma)$ by the term $\frac{\partial}{\partial\nu}\,(\Delta\phi) \in L_2(\Sigma)$ (see comments below (7.44)). This is done in [Hor.1] using the idea of [Leb.1]. However, additional difficulties arise because of the inclusion of the boundary operator B_1 in (7.87) which gives rise to the non-homogeneous right-hand side in equation (7.96). To bound the additional terms which arise from the right-hand side of (7.96), a regularity result for a specific Schrödinger equation must be established. Exact controllability on [0,T] with just one control $u \in L_2(\Sigma)$ can then be shown to hold on the space (7.93) as desired. Thus, the required Finite Cost Condition (1.9) does hold true for problem (7.87). Both the idea in [Leb.1] and the regularity estimate for the Schrödinger equation are also used in the proof of the corresponding uniform stabilization result for problem (7.87) obtained in [Hor.2], following [Las.7] for the case $\mu = 1$, i.e., when the boundary operator B_1 is absent.

7.7. Class (H.2): Wave equation with interior point control

We consider the following interior point control problem for the wave equation

$$
\begin{cases}
w_{tt} = \Delta w + \delta(x)u(t) & \text{in } (0,T]\times\Omega = Q; & (7.101a) \\
w(0,\cdot) = w_0; \; w_t(0,\cdot) = w_1 & \text{in } \Omega; & (7.101b) \\
w|_\Sigma \equiv 0 & \text{in } (0,T]\times\Gamma = \Sigma, & (7.101c)
\end{cases}
$$

where $\delta(x)$ is the Dirac mass $+1$ at the point O (origin), assumed to be an interior point of the open bounded domain $\Omega \subset R^n$, $n = 1,2,3$. The control u is assumed in $L_2(0,T)$.

Non-smoothing observation R. Consistently with established (optimal) regularity theory, described below in (7.111), we take

$$
\{w_0, w_1\} \in Y = Y_1 \times Y_2 \overset{\text{def}}{\equiv}
\begin{cases}
L_2(\Omega) \times H^{-1}(\Omega) & n = 3; & (7.102a) \\
H_{00}^{\frac{1}{2}}(\Omega) \times [H_{00}^{\frac{1}{2}}(\Omega)]' & n = 2; & (7.102b) \\
H_0^1(\Omega) \times L_2(\Omega) & n = 1, & (7.102c)
\end{cases}
$$

and, initially, the cost functional

$$
J(u,w) = \int_0^T \{\|w(t)\|_{Y_1}^2 + \|w_t(t)\|_{Y_2}^2 + |u(t)|^2\}dt, \qquad (7.103)
$$

where we recall [T.10] that

$$
H_0^1(\Omega) = \mathcal{D}(A^{\frac{1}{2}}); \quad H_{00}^{\frac{1}{2}}(\Omega) = \mathcal{D}(A^{\frac{1}{4}}) \qquad (7.104)
$$

for the positive self-adjoint operator $Ah = -\Delta h$, $\mathcal{D}(A) = H^2(\Omega) \cap H_0^1(\Omega)$.

Abstract setting. To put problem (7.101)–(7.103) into the abstract model (1.1), (1.2), we take

$$
A = \begin{vmatrix} 0 & I \\ -A & 0 \end{vmatrix}; \quad B = \begin{vmatrix} 0 \\ \delta \end{vmatrix}; \quad R = I; \qquad (7.105)
$$

$$
Z = Y \text{ (as defined in (7.102))}; \quad U = \mathbb{R}^1. \qquad (7.106)
$$

Assumption (1.3). $(-A)^{-\gamma}B \in \mathcal{L}(U,Y)$. As in (7.6), we compute via (7.105) with $u \in \mathbb{R}^1$:

$$(-A)^{-1}Bu = \begin{vmatrix} 0 & A^{-1} \\ -I & 0 \end{vmatrix} \begin{vmatrix} 0 \\ \delta \end{vmatrix} = \begin{vmatrix} A^{-1}\delta \\ 0 \end{vmatrix}, \qquad (7.107)$$

which we shall show to belong to Y. To this end, we recall that for the second-order differential operator A we have $\mathcal{D}(A^{\alpha/2}) \subset H^{\alpha}(\Omega)$, hence,

$$\delta \in [H^{\alpha}(\Omega)]' \subset [\mathcal{D}(A^{\alpha/2})]' \quad \text{or} \quad A^{-\alpha/2}\delta \in L_2(\Omega), \qquad (7.108)$$

after recalling Sobolev embedding, where with $\varepsilon > 0$, $\alpha = \frac{1}{2}+\varepsilon$ for $n = 3$; $\alpha = 1+\varepsilon$ for $n = 2$; $\alpha = \frac{1}{2}+\varepsilon$ for $n = 1$. Thus with $z = A^{-\alpha/2}\delta \in L_2(\Omega)$ we have by (7.109),

$$A^{-1}\delta = A^{\alpha/2\ -1}z$$

$$= \begin{cases} A^{-\frac{1}{4}\ +\varepsilon/2}z \in \mathcal{D}(A^{\frac{1}{4}\ -\varepsilon/2}) \subset L_2(\Omega), & n = 3; & (7.110a) \\ A^{-\frac{1}{2}\ +\varepsilon/2}z \in \mathcal{D}(A^{\frac{1}{2}\ -\varepsilon/2}) \subset \mathcal{D}(A^{\frac{1}{4}}) = H^{\frac{1}{2}}_{00}(\Omega), & n = 2; & (7.110b) \\ A^{-\frac{3}{4}\ +\varepsilon/2}z \in \mathcal{D}(A^{\frac{3}{4}\ -\varepsilon/2}) \subset \mathcal{D}(A^{\frac{1}{2}}) = H^1_0(\Omega), & n = 1. & (7.110c) \end{cases}$$

Thus, by (7.107), (7.110), and (7.102), we see that Assumption (1.3) holds true with $\gamma = 1$.

Assumption (H.2) = (1.6). This is equivalent to its dual version (1.12); i.e., in our case, to the statement that for problem (7.101) with $w_0 = w_1 = 0$ we have:

$$L: u \to \{w(t), w_t(t)\} = y(t): \text{ continuous } L_2(0,T) \to C([0,T];Y) \qquad (7.111)$$

with Y in (7.102). Property (7.111) is indeed true: for $n = 3$ several proofs have been given, see [Lio,2, p. 27] (by J. L. Lions; by Y. Meyer; by L. Nirenberg) and [T.10]; for $n = 2, 1$, see [T.10].

We note that the regularity (7.111) is "$\frac{1}{2}+\varepsilon$" sharper (in the space variable) measured in Sobolev space order, than the regularity that one would obtain by using only property (7.108).

Conclusion: $T < \infty$ (non-smoothing R). On the basis of the foregoing analysis, we obtain that Theorem 3.1 *on the pointwise synthesis of the optimal pair holds true for problem* (7.101)-(7.103).

Smoothing observation R. We now turn to the applicability of Theorem 3.3 on the existence and uniqueness of the Riccati operator under (possibly) smoothing action of the observation operator $R \in \mathcal{L}(Y,Z)$ and final state observation on $G \in \mathcal{L}(Y;W)$ in the cost (1.2), with $y(t) = [w(t), w_t(t)]$.

Assumption (A.1) = (3.25) on R. If $C(t)$ is the cosine operator on $L_2(\Omega)$ generated by A and $S(t) = \int_0^t C(\tau)d\tau$ is the corresponding sine operator: continuous $L_2(\Omega) \to C([0,T];\mathcal{D}(A^{\frac{1}{2}}))$, we compute preliminarily with $u \in \mathbb{R}^1$, via (7.105),

$$
e^{At}Bu = \begin{vmatrix} C(t) & S(t) \\ -AS(t) & C(t) \end{vmatrix} \begin{vmatrix} 0 \\ \delta u \end{vmatrix} = \begin{vmatrix} S(t)\delta u \\ C(t)\delta u \end{vmatrix} = \begin{vmatrix} A^{\frac{\alpha-1}{2}} A^{\frac{1}{2}} S(t)zu \\ A^{\alpha/2} C(t)zu \end{vmatrix} , \quad (7.112)
$$

where $z = A^{-\alpha/2}\delta \in L_2(\Omega)$ by (7.108). Recalling the values of α in (7.109), and the above property of $S(t)$, we obtain from (7.112),

$$
e^{At}Bu \in \begin{cases} C([0,T];[\mathcal{D}(A^{\frac{1}{4}+\varepsilon/2})]' \times [\mathcal{D}(A^{\frac{1}{4}+\varepsilon/2})]'), & n = 3; \quad (7.113a) \\ C([0,T];[\mathcal{D}(A^{\varepsilon/2})]' \times [\mathcal{D}(A^{\frac{1}{2}+\varepsilon/2})]'), & n = 2; \quad (7.113b) \\ C([0,T];\mathcal{D}(A^{\frac{1}{4}-\varepsilon/2}) \times [\mathcal{D}(A^{\frac{1}{4}+\varepsilon/2})]'), & n = 1. \quad (7.113c) \end{cases}
$$

Thus, by (7.113) we see that *a-fortiori* Assumption (A.1) = (3.25) is satisfied provided that the observation operator $R \in \mathcal{L}(Y,Z)$, Y as in (7.102), satisfies

R^*R: continuous

$$
\begin{cases} [\mathcal{D}(A^{\frac{1}{4}+\varepsilon/2})]' \times [\mathcal{D}(A^{\frac{1}{4}+\varepsilon/2})]' \\ [\mathcal{D}(A^{\varepsilon/2})]' \times [\mathcal{D}(A^{\frac{1}{2}+\varepsilon/2})]' \\ [\mathcal{D}(A^{\frac{1}{4}-\varepsilon/2}) \times [\mathcal{D}(A^{\frac{1}{4}+\varepsilon/2})]' \end{cases} \longrightarrow \begin{cases} L_2(\Omega) \times [\mathcal{D}(A^{\frac{1}{2}})]' & n = 3; \quad (7.114a) \\ \mathcal{D}(A^{\frac{1}{4}}) \times [\mathcal{D}(A^{\frac{1}{4}})]' & n = 2; \quad (7.114b) \\ \mathcal{D}(A^{\frac{1}{2}}) \times L_2(\Omega) & n = 1, \quad (7.114c) \end{cases}
$$

which requires that R^*R be smoothing; e.g., $R = A^{-1/8-\varepsilon}$.

Assumption (A.2) = (3.26) on G. If $g \in L_1(0,T;U)$ arbitrary, we need to require in order to satisfy (A.2) = (3.26), that the final state observation $G \in \mathcal{L}(Y,W)$ makes

$$\int_0^T (B^* e^{A^* t} G^* x, \ g(t))_U dt \ = \ \int_0^T (x, Ge^{At} Bg(t))_Y dt \qquad (7.115)$$

well defined, for $x \in Y$. But $e^{At} Bg(t)$ has the regularity expressed by (7.113) where $C([0,T])$ there is replaced by $L_1(0,T)$ now. Thus, G must have the same (smoothing) properties as the operator $R^* R$, as described by (7.114); e.g., $G = A^{-\frac{1}{4} - \varepsilon}$.

<u>Conclusion</u>: $T < \infty$ (smoothing R and G). Theorem 3.3 is applicable to the dynamics (7.101) with interior point control on the spaces described above, provided R and G are (smoothing) operators, in the sense that $R^* R$ and G satisfy the lifting property (7.114).

<u>Remark 7.2</u>. A similar analysis holds true for problem (7.101a-b) with Dirichlet B.C. (7.101c) replaced by the Neumann B.C. $\frac{\partial w}{\partial \nu}\big|_\Sigma \equiv 0$: the corresponding regularity results are given in [T.10] and they coincide in terms of spaces based on domains of fractional powers with those of (7.101).

<u>Case $T = \infty$</u>. Exact controllability (hence uniform stabilization) for problem (7.101) with (finitely many) interior point control(s) in $L_2(0,T)$ on the space Y of regularity in (7.102) is not possible [T.8]. Thus, the corresponding problem with cost given by (7.103) with $T = \infty$ cannot admit a Riccati theory.

7.8.   <u>Class (H.2): Kirchhoff equation with interior point control</u>
       We consider the following interior point control problem for the Kirchhoff equation

$$\begin{cases} w_{tt} - \rho \Delta w_{tt} + \Delta^2 w = \delta(x) u(t) & \text{in } (0,T] \times \Omega = Q; & (7.116a) \\ w(0, \cdot) = w_0, \ w_t(0, \cdot) = w_1 & \text{in } \Omega; & (7.116b) \\ w\big|_\Sigma \equiv 0 & \text{in } (0,T] \times \Gamma = \Sigma; & (7.116c) \\ \Delta w\big|_\Sigma \equiv 0 & \text{in } \Sigma, & (7.116d) \end{cases}$$

where as in (7.28a), $\rho > 0$ is a constant, and where δ is the Dirac mass + 1 exercised at the origin, assumed to be an interior point of the open bounded domain $\Omega \subset R^n$, $n = 1, 2, 3$. Again the control u is assumed in $L_2(0,T)$.

Non-smoothing observation R. Consistently with (optimal) regularity theory [T.11], described below in (7.127), we take

$$\{w_0, w_1\} \in Y = Y_1 \times Y_2 \overset{\text{def}}{=} \begin{cases} \mathcal{D}(A^{\frac{1}{2}}) \times \mathcal{D}(A^{\frac{1}{4}}), & n = 3; \quad (7.117a) \\ \mathcal{D}(A^{\frac{3}{4}}) \times \mathcal{D}(A^{\frac{3}{8}}), & n = 2; \quad (7.117b) \\ \mathcal{D}(A^{\frac{3}{4}}) \times \mathcal{D}(A^{\frac{1}{2}}), & n = 1, \quad (7.117c) \end{cases}$$

and, initially, the cost functional

$$J(u,w) = \int_0^T \{\|w(t)\|_{Y_1}^2 + \|w_t(t)\|_{Y_2}^2 + |u(t)|^2\} dt. \tag{7.118}$$

In (7.117) we have that A is the positive self-adjoint operator defined as in Section 7.5, or in (6.27), by

$$Ah = \Delta^2 h; \quad \mathcal{D}(A) = \{h \in H^4(\Omega): h|_\Gamma = \Delta h|_\Gamma = 0\}. \tag{7.119}$$

Then [Gr.1]

$$\mathcal{D}(A^{\frac{3}{4}}) = \{h \in H^3(\Omega): h|_\Gamma = \Delta h|_\Gamma = 0\}; \quad \mathcal{D}(A^{\frac{1}{4}}) = H_0^1(\Omega); \tag{7.120}$$

$$A^{\frac{1}{2}} h = -\Delta h; \quad \mathcal{D}(A^{\frac{1}{2}}) = H^2(\Omega) \cap H_0^1(\Omega). \tag{7.121}$$

Abstract setting. To put problem (7.116)-(7.118) into the abstract model (1.1), (1.2), we first observe via (7.120), (7.121), that the dynamics (7.116) can be rewritten abstractly as

$$(I + \rho A^{\frac{1}{2}}) w_{tt} + Aw = \delta u, \tag{7.122}$$

and then we take as in (7.80),

$$\mathcal{A} = \begin{vmatrix} 0 & I \\ -A & 0 \end{vmatrix}; \quad B = \begin{vmatrix} 0 \\ (I + \rho A^{\frac{1}{2}})^{-1} \delta \end{vmatrix}; \quad A = (I + \rho A^{\frac{1}{2}})^{-1} A; \quad R = I. \tag{7.123}$$

$$Z = Y \text{ (as defined in (7.117))}; \quad U = \mathbb{R}^1.$$

Assumption (1.3). $(-\mathcal{A})^{-\gamma} B \in \mathcal{L}(U, Y)$. With $u \in \mathbb{R}^1$ we compute

$$(-\mathcal{A})^{-1} Bu = \begin{vmatrix} 0 & A^{-1} \\ -I & 0 \end{vmatrix} \begin{vmatrix} 0 \\ (I + \rho A^{\frac{1}{2}})^{-1} \delta u \end{vmatrix} = \begin{vmatrix} A^{-1} \delta u \\ 0 \end{vmatrix}, \tag{7.124}$$

which we shall show to belong to Y. To this end, we recall that for the fourth-order operator A in (7.119) we have $\mathcal{D}(A^{\alpha/4}) \subset H^{\alpha}(\Omega)$, hence

$$\delta \in [H^{\alpha}(\Omega)]' \subset [\mathcal{D}(A^{\alpha/4})]' \quad \text{or} \quad A^{-\alpha/4}\delta \in L_2(\Omega), \qquad (7.125)$$

where α takes on the values described in (7.109). Thus with $z = A^{-\alpha/4}\delta \in L_2(\Omega)$, we have by (7.109),

$$A^{-1}\delta = A^{\alpha/4 - 1}z = \begin{cases} A^{-5/8 + \varepsilon/4}z \in \mathcal{D}(A^{5/8 - \varepsilon/4}) \subset \mathcal{D}(A^{1/2}), & n = 3; \quad (7.126a) \\ A^{-3/4 + \varepsilon/4}z \in \mathcal{D}(A^{3/4 - \varepsilon/4}) \subset \mathcal{D}(A^{5/8}), & n = 2; \quad (7.126b) \\ A^{-7/8 + \varepsilon/4}z \in \mathcal{D}(A^{7/8 - \varepsilon/4}) \subset \mathcal{D}(A^{3/4}), & n = 1. \quad (7.126c) \end{cases}$$

Thus by (7.124), (7.126), and (7.117), we see that assumption (1.3) holds true with $\gamma = 1$.

Assumption (H.2) = (1.6). This is equivalent to its dual version (1.12); i.e., in our case, to the statement that for problem (7.116) with $w_0 = w_1 = 0$ we have

$$L: u \to \{w(t), w_t(t)\} = y(t): \text{ continuous } L_2(0,T) \to C([0,T];Y) \quad (7.127)$$

with Y as in (7.117). The validity of the regularity property (7.127) has been provided in [T.11]. As in the case of problem (7.101) of the preceding section, we note that the regularity (7.127) is "$1/2+\varepsilon$" sharper (in the space variable) measured in Sobolev space order, than the regularity that one would obtain by using only property (7.125).

Conclusion: $T < \infty$ (non-smoothing R). On the basis of the foregoing analysis, we obtain that Theorem 3.1 *on the pointwise synthesis of the optimal pair holds true for problem* (7.116)-(7.118).

Smoothing observation operators R and G. We now turn to the applicability of Theorem 3.3 on the existence and uniqueness of the Riccati operator under (possibly) smoothing action of the observation operator $R \in \mathcal{L}(Y,Z)$ and final state observation $G \in \mathcal{L}(Y,W)$ in the cost (1.2), with $y(t) = [w(t), w_t(t)]$.

Assumption (A.1) = (3.25) on R. The operator [L-T.16; App. C],

$$-A = -(I+\rho A^{\frac{1}{2}})^{-1}A = \frac{-A^{\frac{1}{2}}}{\rho} + \frac{I}{\rho^2} - \frac{1}{\rho^2}(I+\rho A^{\frac{1}{2}})^{-1} \qquad (7.128)$$

a bounded perturbation of $-A^{\frac{1}{2}}/\rho$, generates a s.c. cosine operator $\mathcal{C}(t)$ on $L_2(\Omega)$ with $\mathcal{S}(t) = \int_0^t \mathcal{C}(\tau)d\tau$: continuous $L_2(\Omega) \to C([0,T];\mathcal{D}(A^{\frac{1}{2}}))$. From (7.122), (7.123), we then obtain with $u \in \mathbb{R}$

$$e^{At}Bu = \begin{bmatrix} \mathcal{C}(t) & \mathcal{S}(t) \\ -A\mathcal{S}(t) & \mathcal{C}(t) \end{bmatrix} \begin{bmatrix} 0 \\ (I+\rho A^{\frac{1}{2}})^{-1}\mathcal{6} \end{bmatrix}$$

$$= \begin{vmatrix} A^{\frac{\alpha-1}{4}}(I+\rho A^{\frac{1}{2}})^{-1}A^{\frac{1}{4}}\mathcal{S}(t)zu \\ A^{\alpha/4}(I+\rho A^{\frac{1}{2}})^{-1}\mathcal{C}(t)zu \end{vmatrix}, \qquad (7.129)$$

where $z = A^{-\alpha/4}\mathcal{6} \in L_2(\Omega)$ by (7.125). Recalling the values of α in (7.109) and the above property of $\mathcal{S}(t)$ we obtain from (7.129),

$$e^{At}Bu \in \begin{cases} C([0,T];\mathcal{D}(A^{\frac{2}{3}-\varepsilon/4}) \times \mathcal{D}(A^{\frac{1}{3}-\varepsilon/4})), & n = 3; \qquad (7.130a) \\ C([0,T];\mathcal{D}(A^{\frac{1}{2}-\varepsilon/4}) \times \mathcal{D}(A^{\frac{1}{4}-\varepsilon/4})), & n = 2; \qquad (7.130b) \\ C([0,T];\mathcal{D}(A^{\frac{2}{3}-\varepsilon/4}) \times \mathcal{D}(A^{\frac{2}{3}-\varepsilon/4})), & n = 1. \qquad (7.130c) \end{cases}$$

Thus, by (7.130) we see that *a-fortiori* assumption (A.1) = (3.25) is satisfied provided that the observation operator $R \in \mathcal{L}(Y,Z)$, Y as in (7.117), satisfies

R^*R: continuous

$$\begin{cases} \mathcal{D}(A^{\frac{2}{3}-\varepsilon/4}) \times \mathcal{D}(A^{\frac{1}{3}-\varepsilon/4}) \\ \mathcal{D}(A^{\frac{1}{2}-\varepsilon/4}) \times \mathcal{D}(A^{\frac{1}{4}-\varepsilon/4}) \\ \mathcal{D}(A^{\frac{5}{3}-\varepsilon/4}) \times \mathcal{D}(A^{\frac{3}{3}-\varepsilon/4}) \end{cases} \longrightarrow \begin{cases} \mathcal{D}(A^{\frac{1}{2}}) \times \mathcal{D}(A^{\frac{1}{4}}), & n = 3; \quad (7.131a) \\ \mathcal{D}(A^{\frac{2}{3}}) \times \mathcal{D}(A^{\frac{1}{3}}), & n = 2; \quad (7.131b) \\ \mathcal{D}(A^{\frac{2}{4}}) \times \mathcal{D}(A^{\frac{1}{2}}), & n = 1; \quad (7.131c) \end{cases}$$

which requires that R^*R be smoothing; e.g., $R = A^{-1/16 -\varepsilon}$.

Assumption (A.2) = (3.26) on G. As in the preceding section (see (7.115)), we obtain that G must have the same smoothing properties on the operator R^*R, as described by (7.131); e.g., $G = A^{-\frac{1}{4}-\varepsilon}$.

Conclusion: $T < \infty$ (smoothing R and G). Theorem 3.3 is applicable to the dynamics (7.116) with interior point control on the spaces identified above provided R and G are smoothing operators, in the sense that R^*R and G satisfy the lifting property (7.131).

Remark 7.3. If the B.C. $\Delta w|_\Sigma \equiv 0$ in (7.116c) is replaced by $\frac{\partial w}{\partial \nu}|_\Sigma \equiv 0$, then the corresponding Kirchhoff problem has the same regularity property (7.127), (7.117), in terms of the spaces of fractional powers of the new operator A which incorporates the new B.C. This is proved in [T.11] to which we refer for the new technical issues that appear in this case.

Case $T = \infty$. As in the previous subsection, exact controllability (hence uniform stabilization) for problem (7.116) with (finitely many) interior point controls in $L_2(0,T)$ is not possible on the space Y of regularity in (7.117) [T.8].

Remark 7.4. Similar analyses work also in the following two cases: (i) Euler-Bernoulli equations with interior point control, and (ii) Schrödinger equations with interior point controls. Their sharp regularity theory is provided in [T.10] and [T.12]. All the considerations/conclusions of Sections 7.7 and 7.8 hold true *mutatis mutandis*.

8. **Example of a partial differential equation problem satisfying (H.2$_R$)**

   The present section serves as an illustration of both Theorem 4.1 and 4.2 by means of a boundary control/boundary observation problem for hyperbolic mixed problems of Neumann type. This is, in fact, a key example which motivated the introduction of the class (H.2$_R$) = (1.8).

All assumptions of Theorem 4.1 and 4.2 will be shown to be satisfied in this case. However, as we shall see, this task will *critically* rely on the recent sharp regularity theory of second-order hyperbolic equations of Neumann type [L-T.20], [L-T.23], while earlier theory [L-M] and [M.1] will be shown to be insufficient and inadequate.

8.1. **Boundary control/boundary observation for hyperbolic mixed problems of Neumann type. Application of Theorems 4.1 and 4.2**

   With Ω an open bounded domain in R^n, $n \geq 2$, with sufficiently smooth boundary Γ, we consider the following mixed problem of Neumann type,

$$\begin{cases} w_{tt} - \Delta w + w = 0 & \text{(8.1a)} \\ w(0,\cdot) = w_0; \ w_t(0,\cdot) = w_1 & \text{in } \Omega; & \text{(8.1b)} \\ \dfrac{\partial w}{\partial \nu} = u & \text{in } \Sigma = (0,T] \times \Gamma. & \text{(8.1c)} \end{cases}$$

(As noted in Remark 8.1 below, the case dim $\Omega = 1$ is much more regular.) The optimal control problem is now: with $0 < T < \infty$ preassigned, minimize

$$J(u,w) = \int_0^T \{ \|w(t)|_\Gamma\|^2_{L_2(\Gamma)} + \|u\|^2_{L_2(\Gamma)} \}\, dt \qquad (8.2)$$

over all $u \in L_2(0,T;L_2(\Gamma)) = L_2(\Sigma)$, with w solution of (8.1) due to u. We shall show that this optimal control problem is a specialization of the problem considered in Section 4.1 for (1.1), (1.2).

Abstract setting. The abstract setting for the mixed problem (8.1) is as follows. The abstract spaces Y and U of model (1.1), and the observation space Z are

$$Y = H^1(\Omega) \times L_2(\Omega); \quad U = L_2(\Gamma); \quad Z = L_2(\Gamma). \qquad (8.3)$$

The operators A and B of model (1.1) are

$$A = \begin{vmatrix} 0 & I \\ -A & 0 \end{vmatrix}; \quad Bu = \begin{vmatrix} 0 \\ ANu \end{vmatrix}; \quad A^{-1}Bu = \begin{vmatrix} -Nu \\ 0 \end{vmatrix}; \qquad (8.4)$$

$$-Ah = (\Delta - 1); \quad \mathcal{D}(A) = \{ h \in H^2(\Omega) : \frac{\partial h}{\partial \nu}\big|_\Gamma = 0 \}; \qquad (8.5)$$

$$v = Ng \iff \{ (\Delta - 1)v = 0 \text{ in } \Omega; \frac{\partial v}{\partial \nu}\big|_\Gamma = g \}; \qquad (8.6)$$

$$N: H^s(\Gamma) \to H^{s+\frac{1}{2}}(\Omega) \qquad (8.7a)$$

$$L_2(\Omega) \to H^{\frac{1}{2}}(\Omega) \subset H^{\frac{1}{2}-2\rho}(\Omega) = \mathcal{D}(A^{\frac{1}{4}-\rho}) \qquad (8.7b)$$

continuously. Finally, the observation operator R is:

$$R: Y \supset \mathcal{D}(R) \to Z = L_2(\Gamma): R\begin{vmatrix} y_1 \\ y_2 \end{vmatrix} = y_1|_\Gamma = N^*Ay_1. \qquad (8.8)$$

Remark 8.1. When dim $\Omega = 1$, the situation drastically simplifies. With reference to problem (8.1), elementary methods (including explicit solution formulas) yield

$$L: u \in L_2(\Sigma) \rightarrow \{w, w_t\} \in C([0,T]; H^1(\Omega) \times L_2(\Omega)), \qquad (8.9)$$

$$RL: u \in L_2(\Sigma) \rightarrow w|_\Sigma \in H^1(0,T;R^2), \qquad (8.10)$$

while these results are definitely false for dim $\Omega \geq 2$ [L-T.21]. For dim $\Omega = 1$, the setting of Section 3.3, in particular Theorem 3.3 yielding existence and uniqueness, is applicable. ∎

Remark 8.2. (Sharp regularity theory of problem (8.1)) While we refer to [L-T.20], [L-T.23] for further results and proofs, we quote here only a few results when dim $\Omega \geq 2$:

$$L: u \in L_2(\Sigma) \rightarrow \{w, w_t\} \in C([0,T]; H^\alpha(\Omega) \times H^{\alpha-1}(\Omega)), \quad w_0 = w_1 = 0, \quad (8.11)$$

$$u = 0; \{w_0, w_1\} \in H^1(\Omega) \times L_2(\Omega) \rightarrow w|_\Sigma \in H^\beta(\Sigma), \qquad (8.12)$$

continuously, where α, β are constants strictly greater than ½:
½ < $\alpha \leq$ ¾ which may depend on the geometry of the smooth bounded domain
Ω; examples: $\alpha = \beta = $ ⅗ for a sphere; $\alpha = \beta = $ ¾-ε for a parallelpiped;
$\alpha = $ ¾-ε; $\beta = $ ⅗ for a general domain. Instead, in earlier regularity
theory as in [L-M.1], [M.1], $\alpha = $ ½ and $\beta = $ ½. ∎

Verification of assumptions. We shall now verify that all required
assumptions (1.3), (H.2$_R$), and (h.0) through (h.5) are verified. In
doing so, we shall point out the several places where these
verifications rely *critically* on the recent sharp regularity theory
briefly recalled in Remark 8.2 above, where $\alpha, \beta > $ ½. Instead, the
earlier theory with $\alpha = \beta = $ ½ would be insufficient. For details we
refer to [L-T.10]. ∎

Verification of (1.3): $(-A)^{-\gamma}B$. This is certainly satisfied with $\gamma = 1$
by (8.4), (8.7).

Verification of (H.2$_R$) = (1.8). Using (8.8) for R, we compute that
$R^*z = [Nz, 0]$, R^* the Y-adjoint of R, see (8.3). Moreover,
$B^*[z_1, z_2] = z_2|_\Gamma = N^*Az_2$. Then we obtain for $z \in Z = L_2(\Gamma)$ and with
$S(t)$ the sine operator associated with A:

$$B^*e^{A^*t}R^*z = AS(t)Nz|_\Sigma \in H^{\alpha-\frac{1}{2}-2\rho+(\beta-\alpha)(\frac{1}{2}-2\rho)}(\Sigma) \qquad (8.13)$$

$$\subset L_2(\Sigma), \qquad (8.14)$$

see e.g., [L-T.20], [L-T.23], [L-T.24], where in going from (8.13) to (8.14) we use crucially that $\beta \geq \alpha > \frac{1}{2}$ (since ρ in (8.7) is arbitrarily small) as in the sharp theory of Remark 8.2, while $\alpha = \beta = \frac{1}{2}$ as in the more classical theory would fail to yield (8.14).

Verification of (h.0) = (4.1). Immediate: R in (8.8) is bounded Y → U.

Verification of (h.1) = (4.2). From (8.8), (8.4), we compute $u \in U = L_2(\Gamma)$

$$Re^{At}Bu = S(t)ANu|_\Sigma \in L_2(\Sigma), \qquad (8.15)$$

and we are in the same situation of (8.13), (8.14).

Verification of (H.2) = (4.3). It is immediate: for $x = [x_1, x_2] \in Y$, see (4.22), and by (4.27),

$$Re^{At}x = [C(t)x_1 + S(t)x_2]|_\Sigma \in C([0,T];H^{\frac{1}{2}}(\Gamma)), \qquad (8.16)$$

with C(t) and S(t) cosine and sine operators associated with A, where now classical interior theory plus standard trace theory are sufficient to produce (8.16), which *a fortiori* verifies (h.2).

Thus, so far, *Theorem 4.1 is applicable to problem* (8.1), (8.2) *with* $Y = H^1(\Omega) \times L_2(\Omega)$. We shall now see that Theorem 4.2 applies as well, with

$$\mathcal{U}_{[t,T]} = H^{\frac{1}{2}}(\Sigma_t) = L_2(t,T;H^{\frac{1}{2}}(\Gamma)) \cap H^{\frac{1}{2}}(t,T;L_2(\Gamma)). \qquad (8.17)$$

Verification of (h.3) = (4.15). With $\mathcal{U}_{[0,T]}$ given by (8.17), more classical regularity theory [M.1] is sufficient to yield (4.15) (while sharp regularity theory [L-T.23] yields an even stronger result).

Verification of (h.4) = (4.16). By (1.10), for L_0^*, R^* above (8.13), and (8.16), we shall show that with $x = [x_1, x_2] \in Y$, we have

$$\{L_0^* R^* [\text{Re}^{A \cdot x}]\}(t) = \phi(t)|_\Sigma \in H^{\frac{1}{2}}(\Sigma), \qquad (8.18)$$

where

$$\begin{cases} \phi_{tt} = \Delta\phi - \phi & \text{in } Q; & (8.19a) \\ \phi(T, \cdot) = \phi_t(T, \cdot) = 0 & \text{in } \Omega; & (8.19b) \\ \dfrac{\partial\phi}{\partial\nu}\Big|_\Sigma = \Psi|_\Sigma & \text{in } \Sigma; & (8.19c) \end{cases}$$

$$\Psi(t) = \Psi(t; \Psi_0, \Psi_1) = \text{Re}^{At} x = C(t)x_1 + S(t)x_2; \quad x_1 = \Psi_0, \ x_2 = \Psi_1. \quad (8.20)$$

Plainly from (8.20) with $x \in Y$, we have $\Psi \in H^1(Q)$ and $\Psi|_\Sigma \in H^{\frac{1}{2}}(\Sigma)$. This used in problem (8.19) gives then $\phi|_\Sigma \in H^{\frac{1}{2}}(\Sigma)$ as desired, just by invoking a more classical result [M.1] (while sharp regularity theory [L-T.23] would yield a stronger result). Thus (8.18) is proved and, by (8.17), (h.4) is verified.

Verification of (h.5) = (4.17). We shall verify (h.5) for $t = 0$. The case $t \neq 0$ is similar. By (1.9), (1.10), we shall show that for $u \in H^{\frac{1}{2}}(\Sigma)$ we have

$$\{L_0^* R^* R L_0 u\}(t) = \varsigma(t)|_\Sigma \in H^{2\alpha - \frac{1}{2}}(\Sigma), \qquad (8.21)$$

where

$$\begin{cases} \varsigma_{tt} = \Delta\varsigma - \varsigma & \text{in } Q; & (8.22a) \\ \varsigma(T, \cdot) = \varsigma_t(T, \cdot) = 0 & \text{in } \Omega; & (8.22b) \\ \dfrac{\partial\varsigma}{\partial\nu}\Big|_\Sigma = w|_\Sigma & \text{in } \Sigma, & (8.22c) \end{cases}$$

and $w(t) = w(t; 0, 0)$ is the solution of (8.1) for $w_0 = w_1 = 0$. By using the more classical result of [M.1] on problem (8.1) with $w_0 = w_1 = 0$ we have:

$$u \in H^{\frac{1}{2}}(\Sigma) \to w|_\Sigma \in H^{\frac{1}{2}}(\Sigma) \qquad (8.23)$$

continuously, while the sharp theory [L-T.20], [L-T.23] yields the stronger result

$$u \in H^{\frac{1}{2}}(\Sigma) \to w|_\Sigma \in H^{2\alpha - \frac{1}{2}}(\Sigma) \qquad (8.24)$$

continuously, where $\alpha > \frac{1}{2}$ is defined in Remark 8.2. At this stage we may still use the conservative regularity (8.23) for $w|_\Sigma$ which enters (8.22c); however, it is at the level of analyzing the resulting regularity of $\varsigma|_\Sigma$ that we *critically* use the counterpart of estimate (8.24) (i.e., the sharp theory):

$$w|_\Sigma \in H^{\frac{1}{2}}(\Sigma) \to \varsigma|_\Sigma \in H^{2\alpha-\frac{1}{2}}(\Sigma), \qquad (8.25)$$

and (8.21) follows. But $2\alpha-\frac{1}{2} > \frac{1}{2}$, see Remark 8.2; thus

$$\text{the injection } H^{2\alpha-\frac{1}{2}}(\Sigma) \to H^{\frac{1}{2}}(\Sigma) \text{ is compact.} \qquad (8.26)$$

Putting together (8.21), (8.25), and (8.26), we obtain

$$L_0^* R^* R L_0: \text{ compact } H^{\frac{1}{2}}(\Sigma) \to H^{\frac{1}{2}}(\Sigma) \qquad (8.27)$$

as desired, and (h.5) = (4.17) is verified. Conculsion (8.27) would not follow instead, using the earlier theory $\alpha = \frac{1}{2}$.

<u>Conclusion</u>. We have verified all the required assumptions (1.3), $(H.2_R)$, (h.0) through (h.5) for problem (8.1), (8.2). Thus, both Theorems 4.1 and 4.2 are applicable to this problem. We obtain the following specialization:

<u>Theorem 8.1</u>. With reference to the optimal boundary control/boundary observation problem (8.1), (8.2), we have:

(i)   there exists a unique solution of the DRE, $\forall \, x, z \in H^1(\Omega) \times L_2(\Omega)$,

$$\frac{d}{dt} (P(t)x,z)_{H^1(\Omega) \times L_2(\Omega)} = (x_1|_\Gamma, z_1|_\Gamma)_{L_2(\Gamma)} - (P(t)x, Az)_{H^1(\Omega) \times L_2(\Omega)}$$

$$- (P(t)Ax, z)_{H^1(\Omega) \times L_2(\Omega)} + ([P(t)x]_2|_\Gamma, [P(t)z]_2|_\Gamma)_{L_2(\Gamma)} \qquad (8.28)$$

with $P(T) = 0$, where we write $P(t)x = \{[P(t)x]_1, [P(t)x]_2\}$ for the two components in $H^1(\Omega) \times L_2(\Omega)$.

(ii)   Uniqueness is within the class of the following properties:

(ii_1) $P(t) = P^*(t) \geq 0, \quad 0 \leq t \leq T$ (* in $H^1(\Omega) \times L_2(\Omega)$); (8.29)

(ii_2) $P(t) \in \mathcal{L}(H^1(\Omega) \times L_2(\Omega); C([0,T]; H^1(\Omega) \times L_2(\Omega));$ (8.30)

(ii_3) $\| [P(t)x]|_\Gamma \|_{C([0,T]; L_2(\Gamma))} \leq c_T \|x\|_{H^1(\Omega) \times L_2(\Omega)}$.   (8.31)

(iii) The pointwise feedback representation of the unique optimal pair for the problem starting at t = 0 is:

$$u^0\left(t; \begin{vmatrix} w_0 \\ w_1 \end{vmatrix}\right) = -\left[P(t) \begin{vmatrix} w^0(t; w_0, w_1) \\ w^0_t(t; w_0, w_1) \end{vmatrix}\right]_2\Bigg|_\Gamma . \tag{8.32}$$

(iv) The optimal cost is

$$J\left(u^0\left(\cdot; \begin{vmatrix} w_0 \\ w_1 \end{vmatrix}\right), \begin{vmatrix} w^0(\cdot; w_0, w_1) \\ w^0_t(\cdot; w_0, w_1) \end{vmatrix}\right) = (P(0)\begin{vmatrix} w_0 \\ w_1 \end{vmatrix}, \begin{vmatrix} w_0 \\ w_1 \end{vmatrix})_{H^1(\Omega) \times L_2(\Omega)} . \tag{8.33}$$

(v) The optimal pair satisfies the regularity properties:

$$\sup_{0 \leq t \leq T} |u^0(\cdot, t; x)|_{H^{1/2}(\Sigma_t)} \leq c_T |x|_{H^1(\Omega) \times L_2(\Omega)} ; \tag{8.34}$$

$$\sup_{0 \leq t \leq T} |y^0(\cdot, t; x)|_{C([t,T]; H^1(\Omega) \times L_2(\Omega))} \leq c_T |x|_{H^1(\Omega) \times L_2(\Omega)} \tag{8.35}$$

where $x = \{w_0, w_1\}$, $y^0 = \{w^0, w^0_t\}$. ∎

Remark 8.3. We can also use the setting of Section 3.3 to treat problem (8.1), (8.2); in particular, we can apply Theorem 3.3 to this problem. However, in order to do so, we must take now

$$Y = H^\alpha(\Omega) \times H^{\alpha-1}(\Omega) = \mathcal{D}(A^{\alpha/2}) \times [\mathcal{D}(A^{\frac{1-\alpha}{2}})]' \tag{8.36}$$

instead of the smoother space $Y = H^1(\Omega) \times L_2(\Omega)$ as in (8.3). To this end, all we need is the following:

Verification of assumption (A.1) = (3.25) with U = Z = $L_2(\Gamma)$ and Y as in (8.36). Since $\alpha > \frac{1}{2}$ in the sharp theory (Remark 8.2), we have by trace theory that the operator R (Dirichlet trace) in (8.8) satisfies $R \in \mathcal{L}(Y; Z)$. Moreover, $Re^{At}Bu \in L_2(\Sigma) = L_2(0,T;Z)$ with $u \in U = L_2(\Gamma)$ by (8.15). This, combined with $R^* \in L(Z;Y)$, shows (3.25) as desired. ∎

II. Approximation theory

9. Numerical approximations of the solution to the abstract Differential and Algebraic Riccati Equations

The main goal of this section is twofold: (i) to formulate a numerical algorithm for the computation of the solution to the

Differential and Algebraic Riccati Equations (DRE) (3.25) and (ARE) (5.1); (ii) to present the relevant convergence results.

To begin with, we introduce a family of approximating subspaces $V_h \subset Y \cap \mathcal{D}(B^*)$, where h, $0 < h \leq h_0 < \infty$, is a parameter of discretization which tends to zero. Let Π_h be the orthogonal projection of Y onto V_h, with the usual approximating property

$$\|\Pi_h y - y\|_Y \to 0, \quad y \in Y. \tag{9.1}$$

Let $A_h: V_h \to V_h$ and $B_h: U \to V_h$ be approximations of A, respectively B, which satisfy the usual, natural requirements:

(i) $\qquad\qquad\qquad \Pi_h \hat{A}^{-1} - \hat{A}_h^{-1} \Pi_h \to 0$, strongly in Y; $\qquad$ (9.2a)

(ii) $\qquad\qquad\qquad \|\hat{A}^{-1}(B_h - B)u\|_Y \to 0$, $u \in U$. $\qquad\qquad$ (9.2b)

We consider the following approximation of the DRE (3.25) and ARE (5.1):

$$(\dot{P}_h(t)x_h, y_h)_Y + (A_h^* P_h(t)x_h, y_h)_Y + (P_h(t)A_h x_h, y_h)_Y + (Rx_h, Ry_h)_Z$$

$$= (B_h^* P_h(t)x_h, B_h^* P_h(t)y_h)_U; \qquad\qquad \text{DRE}_h \quad (9.3)$$

$$(P_h(T)x_h, y_h)_Y = (G^* G x_h, y_h), \quad \forall\, x_h, y_h \in V_h;$$

$$(A_h^* P_h x_h, y_h)_Y + (P_h A_h x_h, y_h)_Y + (Rx_h, Ry_h)_Z = (B_h^* P_h x_h, B_h^* P_h y_h)_U ,$$

$$x_h, y_h \in V_h. \qquad (\text{ARE})_h \quad (9.4)$$

Our main goal is to prove that, under natural assumptions which are the discrete counterpart of the hypothesis (H.1) = (1.5) or (H.2) = (1.6) of the continuous case, we have (among other things): in the case of the Differential Riccati Equation

$$P_h(t)\Pi_h x \to P(t)x, \text{ strongly in } C([0,T];Y), \quad x \in Y; \quad (9.5)$$

$$B_h^* P_h(t)\Pi_h x \to B^* P(t)x, \text{ strongly in } C([0,T];U), \quad x \in Y; \quad (9.6)$$

and in the case of the Algebraic Riccati Equation,

(i) $\qquad P_h \to P$, strongly in Y; $\qquad\qquad\qquad\qquad\qquad$ (9.7)

(ii) $B_h^* P_h \to B^* P$, in a technical sense to be made precise;   (9.8)

(iii) $\left\| e^{(A_h - B_h B_h^* P_h) t} \right\|_{\mathcal{L}(Y)} \leq C e^{-\omega t}, \; \omega > 0.$ (9.9)

Although there are a number of papers in the literature which deal with the problem of approximating RE, most of these works [G.1], [B-K.1], [K-S.1], [I-T.1], treat the case where the input operator B is bounded. When instead B is genuinely unbounded, an array of new difficulties arise. Some of them are the same which are already encountered in the continuous case treatment; some others are new, and are intrinsically connected with the approximating schemes. We list a few.

(a) <u>Open loop approximation</u>. Consider the input → solution operator

$$(Lu)(t) = \int_0^t e^{A(t-\tau)} Bu(\tau) d\tau.$$ (9.10)

Under either hypothesis (H.1) = (1.5), or else hypothesis (H.2), the operator L is continuous: $L_2(0,T;U) \to L_2(0,T;Y)$ (indeed → $C([0,T];Y)$ in the case of assumption (H.2), and also in the case of assumption (H.1) with $\gamma < \frac{1}{2}$, or with $\gamma = \frac{1}{2}$, when A has a Riesz basis on Y). In the corresponding approximation theory, the question arises whether the discrete map

$$(L_h u)(t) = \int_0^t e^{A_h(t-\tau)} B_h u(\tau) d\tau$$ (9.11)

which is continuous: $L_2(0,T;U) \to L_2(0,T;V_h)$, is also continuous as an operator $L_2(0,T;U) \to L_2(0,T;Y)$, uniformly with respect to the parameter h. (For instance, one may take $B_h = \Pi_h B$, where we note that $\Pi_h B$ is well defined since $V_h \in \mathcal{D}(B^*)$ by assumption:

$$(\Pi_h Bu, v_h)_Y = (Bu, v_h)_Y = (u, B^* v_h)_U.$$ (9.12)

In the case where B is bounded, this stability requirement is true if the approximation of A is consistent, i.e., subject to (9.2), as it follows via Trotter-Kato theorem. Instead, in the case where B is unbounded, special care must be exercised to select a suitable approximation scheme, which guarantees the above stability requirement.

(b) <u>Approximation of gain operators</u> $B^*P_h(t)$, B^*P_h. The problem here is that even if (9.5) (resp. (9.7)) holds true, it is far from clear that (9.6) (resp. (9.8)) will also follow. Thus, special care must be given in selecting the approximating schemes, as to obtain convergence (9.6), (9.8) for the gain operators.

Thus, a theory of approximations in the case where the operator B is genuinely unbounded, such as it arises in boundary control and point control for partial differential equations, offers new challenges which are not present in the B-bounded case. In order to cope with these difficulties, we need--as in the continuous case--distinguish between dynamics which satisfy assumption (H.1) = (1.5) and dynamics which satisfy assumption (H.2) = (1.6).

9.1. Approximation for the (H.1)-class

9.1.1 Approximation assumptions

<u>Approximation of A</u>. Let $A_h: V_h \to V_h$ be an approximation of A which satisfies the following requirements:

(A.1) (uniform analyticity)

$$|A_h e^{A_h t}|_{\mathcal{L}(Y)} \le \frac{C}{t} e^{(\omega+\varepsilon)t}; \quad t > 0 \qquad (9.13)$$

discrete analog of (H.1), where the constant C is uniform with respect to h;

(A.2) $$|\Pi_h \hat{A}^{-1} - \hat{A}_h^{-1} \Pi_h|_{\mathcal{L}(Y)} \le Ch^s \quad \text{for some } s > 0. \qquad (9.14)$$

<u>Approximation of B</u>. We shall assume that the operator $B: U \to [\mathcal{D}(A^*)]'$ and $B_h: U \to V_h$ satisfy the following "approximation" properties, where γ and s were defined above

(A.3) ('inverse approximation property')

$$\|B^*x_h\|_U + \|B_h^*x_h\|_U \le C h^{-\gamma s} \|x_h\|_H, \quad \forall x_h \in V_h; \qquad (9.15)$$

(A.4) $$\|B^*(\Pi_h - I)x\|_U \le C h^{s(1-\gamma)} \|x\|_{\mathcal{D}(A^*)}, \quad x \in \mathcal{D}(A^*); \qquad (9.16)$$

(A.5) $\qquad \|B^*x - B_h^*\overline{\Pi}_h x\|_U \leq C\, h^{s(1-\gamma)}\|x\|_{\mathcal{D}(A^*)}$, $x \in \mathcal{D}(A^*)$; $\qquad$ (9.17)

(If, in particular, we take $B_h = \overline{\Pi}_h B$, then (A.5) is contained in (A.4).)

(A.6) $\qquad \|B^*\overline{\Pi}_h x\|_U \leq C\|(\hat{A}^*)^\gamma x\|_H$, $x \in \mathcal{D}((\hat{A}^*)^\gamma)$. $\qquad$ (9.18)

Remark 9.1. Notice that (A.2) throughout (A.4) are the standard approximation properties. They are consistent with the regularity of the original operators A and B. Moreover, they are satisfied by typical schemes (finite elements, finite differences, mixed methods, spectral approximations). The property of uniform analyticity (A.1) is not a standard assumption and needs to be verified in each case. However, to our knowledge, it is satisfied for most of the schemes and examples which arise from analytic semigroup problems. For instance, sufficient condition for (A.1) to hold is the uniform coercitivity of the bilinear form associated with A_h (see Lemma 4.2 in [Las.1]). There are, however, a number of significant physical examples (e.g., damped elastic systems) where the bilinear form is not coercive, while the underlying semigroups $e^{A_h t}$ are uniformly analytic (see Section 10).

9.1.2 Consequences of approximating assumptions on A and B

From (A.2) and (A.1), the following "rough" data estimates follow (see [Las.1, Appendix]).

(i) $\qquad |e^{A_h t}\overline{\Pi}_h - \overline{\Pi}_h e^{At}|_{\mathcal{L}(Y)} \leq \dfrac{Ch^{s\Theta}\, e^{(\omega+\varepsilon)t}}{t^\Theta}$, $\qquad$ (9.19)

where $0 \leq \Theta \leq 1$ and $\varepsilon > 0$ can be arbitrarily small;

(ii) $\qquad |\overline{\Pi}_h R(\lambda, A) - R(\lambda, A_h)\overline{\Pi}_h|_{\mathcal{L}(Y)} \leq C\, h^s$, $s > 0$ $\qquad$ (9.20)

uniformly in $\lambda \in \Sigma_{app}^{co}(A)$, where $\Sigma_{app}(A)$ = closed triangular sector containing the axis $[-\infty, a]$ and delineated by the two rays $a + \rho^{\pm i\phi}$ for some $\pi/2 < \phi < 2\pi$; $a = \omega+\varepsilon$;

(iii) $\qquad |e^{A_h^* t}\overline{\Pi}_h - \overline{\Pi}_h e^{A^* t}|_{\mathcal{L}(\mathcal{D}(A^*), Y)} \leq C\, h^s$ $\qquad$ (9.21)

uniformly in $t > 0$ on compact subintervals.

<u>Remark 9.2</u>. We consider the special case of coercive bilinear forms, and show that in this case assumption (A.1) is automatically satisfied. With Y the given Hilbert space, let W be another Hilbert space for which the identity W $\hookrightarrow$ Y is continuous. Assume that the operator A satisfies further the following conditions:

(i) continuity of sesquilinear form on W: there exists a constant K such that

$$|(Ax,y)_Y| \leq K\|x\|_W\|y\|_W \quad \forall x,y \in W;$$

(ii). (Gärding inequality) there exist positive constants c_1, c_2 such that

$$\text{Re}(-Ax,x)_Y \geq c_1\|x\|_W^2 - c_2\|x\|_Y^2, \quad \forall x \in W,$$

so that $-A+c_2 I$ is W-elliptic or coercive.

Then, as is well known, e.g. [Sh., p. 99], the operator A actually generates an analytic semigroup on Y.

With V_h the approximating subspaces introduced before, define $A_h: V_h \to V_h$ by

$$(A_h x_h, y_h)_Y = (Ax_h, y_h)_Y, \quad x_h, y_h \in V.$$

Then, A_h satisfies automatically the continuity and Gärding conditions (i) and (ii) above with the same constants K, c_1, c_2, independently of h > 0. Therefore, the very same argument which proves analyticity of e^{At} in the continuous case, once applied (essentially verbatim) to the discrete case, yields that $e^{A_h t}$ satisfies the uniform analyticity condition (A.1), with constant C independent of h. ∎

9.1.3. <u>Approximation of dynamics and of control problems. Related Riccati equation</u>

We now introduce an approximation of the control problem and of the corresponding RE.

<u>Control problem</u>. Given the approximating dynamics $y_h(t) \subset V_h$ such that

$$\dot{y}_h(t) = A_h y_h(t)+B_h u(t); \quad y(0) = \Pi_h y_0 \tag{9.22}$$

minimize

$$J(u,y_h(u)) \equiv \int_0^T [|Ry_h(t)|_Z^2+|u(t)|_U^2]dt. \tag{9.23}$$

where $T < \infty$ in the case of the DRE and $T = \infty$ in case of the ARE. The optimal solution to (9.22), (9.23) (which we shall see later to exist) will be denoted by $\{u_h^0 = u(y_h^0); y_h^0\}$.

Riccati Equation. The approximation of the Differential/Algebraic Riccati Equation is defined by equation $(\text{DRE}_h) = (9.3)$, $(\text{ARE})_h = (9.4)$ with B_h.

9.1.4. Main results of approximating schemes

Differential Riccati Equation. The main result is

Theorem 9.1. (DRE) Assume hypothesis (H.1) = (1.5) and moreover that $G^*G \in \mathcal{L}(Y; \mathcal{D}(A^*))$. In addition, let the approximating properties (9.1) and (A.1) = (9.13) through (A.6) = (9.18) hold true.

Then there exists $h_0 > 0$ such that for all $0 < h < h_0$, the solution $P_h(t)$ of the $\text{DRE}_h = (9.3)$ exists, is unique, and satisfies the following properties as $h \downarrow 0$:

$$|(\hat{A}_h^*)^{1-\rho} P_h(t)|_{C([0,T];Y)} + |(\hat{A}_h^*)^{\frac{1}{2}-\rho} P_h(t) \hat{A}_h^{\frac{1}{2}-\rho}|_{C([0,T];Y)} \leq c_T,$$

$$\forall \, 0 < \rho \leq 1 ; \quad (9.24)$$

$$|[P_h(t) \Pi_h - P(t)] x|_{C([0,T];Y)} \to 0, \quad x \in Y; \quad (9.25)$$

$$|[B^* P(t) - B_h^* P_h(t) \Pi_h] x|_{C([0,T];U)} \to 0, \quad x \in Y; \quad (9.26)$$

$$|u_h^0 - u^0|_{C([0,T];U)} + |y_h^0 - y^0|_{C([0,T];Y)} + |J(u_h^0, y_h^0) - J(u^0, y^0)| \to 0. \quad (9.27)$$

If, in addition, $V_h \subset \mathcal{D}(\hat{A}^\alpha)$ and the following norm equivalence holds

$$|\hat{A}^\alpha x_h| \sim |A_h^\alpha x_h|, \quad (9.28)$$

then

$$|\hat{A}^{*\alpha} [P_h(t) \Pi_h - P(t)] x|_{C([0,T];Y)} \to 0, \quad x \in Y. \quad \blacksquare \quad (9.29)$$

Algebraic Riccati Equation. In order to obtain approximation results in the case of the ARE, it is necessary to impose some approximation conditions which guarantee that the Finite Cost Condition (1.9) is satisfied by the approximating problem (notice that this does not automatically follow from the fact that the F.C.C. holds true for the continuous problem). Below we impose conditions which are only sufficient, but which are satisfied by all analytic examples to be considered in Section 10.

For the present 'analytic' class, subject to assumption (H.1) = (1.5), the Finite Cost Condition will be guaranteed by the following Stabilizability Condition (S.C.)

$$(S.C.) \quad \begin{cases} \exists \ F \in \mathcal{L}(Y,U) \text{ such that the s.c. analytic semigroup} \\ e^{(A+BF)t} \text{ (as guaranteed by (1.3)) is exponentially} \\ \text{stable on } Y: \\ \quad \| e^{(A+BF)t} \|_{\mathcal{L}(Y)} \leq M_F \ e^{-\omega_F t} \quad \text{for some } \omega_F > 0. \end{cases} \quad (9.30)$$

Our main results are formulated in the theorems below.

Theorem 9.2. (ARE) [L-T.2], [L-T.19] Assume:

I. The continuous hypothesis (H.1) = (1.5), the above Stabilization Condition (9.30), the Detectability Condition (5.10), and, in addition

$$\begin{cases} (i) & \text{either } R > 0 \\ (ii) & \text{or } \hat{A}^{-1}KR: \ Y \to Y \text{ compact}; \end{cases} \quad (9.31)$$

$$\begin{cases} (i) & \text{either } B^*\hat{A}^{*-1}: \ Y \to U \text{ compact} \\ (ii) & \text{or } F: \ Y \to U \text{ compact}. \end{cases} \quad (9.32)$$

II. The approximation properties (9.1), (A.1) = (9.13) through (A.6) = (9.18). Then there exists $h_0 > 0$ such that for all $h < h_0$, the solution P_h to the equation (ARE)$_h$ = (9.4) exists, is unique and the following convergence properties hold:

$$\| e^{-A_{h,p}t} \|_{\mathcal{L}(Y)} \leq C \ e^{-\bar{\omega}_p t}, \quad \bar{\omega}_p > 0, \quad (9.33)$$

where $A_{h,p} \equiv A_h - \Pi_h BB^* P_h$;

$$\| \hat{A}_h^{*1-\rho} P_h \|_{\mathcal{L}(Y)} + \| \hat{A}^{*\frac{1}{2}-\rho} P_h A_h^{\frac{1}{2}-\rho} \|_{\mathcal{L}(Y)} \leq C, \quad \text{for any } 0 < \rho \leq 1; \quad (9.34)$$

$$\| P_h \Pi_h - P \|_{\mathcal{L}(Y)} \leq C \ h^{\varepsilon_0} \to 0 \quad \text{as } h \downarrow 0, \ \forall \ \varepsilon_0 < s(1-\gamma); \quad (9.35)$$

$$\|B_h^* P_h \Pi_h - B^* P\|_{L(Y;U)} \to 0 \quad \text{as } h \downarrow 0; \tag{9.36}$$

for all $\varepsilon_0 < s(1-\gamma)$, as $h \downarrow 0$, $x \in Y$,

$$\sup_{t \geq 0} e^{\overline{\omega}_P t} \|u_h^0(t, \Pi_h x) - u^0(t, x)\|_{\mathcal{L}(Y;U)} \leq C h^{\varepsilon_0} \to 0; \tag{9.38}$$

$$\|y_h^0(\cdot, \Pi_h x) - y^0(\cdot, \Pi x)\|_{\mathcal{L}(Y; L_2(0, \infty; Y))} \leq C h^{\varepsilon_0} \to 0; \tag{9.39}$$

for all $\varepsilon_0 < s(1-\gamma)$ and for all $\varepsilon > 0$, as $h \downarrow 0$,

$$\sup_{t \geq 0} t^\varepsilon e^{\overline{\omega}_P t} \|y_h^0(t, \Pi_h x) - y^0(t, x)\|_{\mathcal{L}(Y)} \leq C h^{\varepsilon_0 \varepsilon} \to 0; \tag{9.40}$$

for all $\varepsilon_0 < s(1-\gamma)$,

$$|J(u_h^0(\cdot, \Pi_h x), y_h^0(\cdot, \Pi_h x)) - J(u^0(\cdot, x), y^0(\cdot, x))| \leq C h^{\varepsilon_0} \to 0. \tag{9.41}$$

Moreover, if in addition, for some $0 < \Theta < 1$, $V_h \subset \mathcal{D}(\hat{A}^\Theta)$ and

$$\|(\hat{A}^*)^\Theta x_h\|_Y \leq C_\Theta \|(\hat{A}_h^*)^\Theta x_h\|_Y, \quad \text{or} \quad (\hat{A}^*)^\Theta (\hat{A}_h^{*-1})^\Theta \in \mathcal{L}(V_h, Y), \tag{9.42}$$

then

$$\|(\hat{A}^*)^\Theta (P_h \Pi_h - P) x\|_Y \to 0 \quad \text{as } h \downarrow 0, \ x \in Y, \ 0 \leq \Theta < 1; \tag{9.43}$$

$$\|(\hat{A}^*)^\Theta (P_h \Pi_h - P) \hat{A}^\Theta x\|_Y \to 0 \quad \text{as } h \downarrow 0, \ x \in Y, \ 0 \leq \Theta < \tfrac{1}{2}. \quad \blacksquare \tag{9.44}$$

Remark 9.3. Assumption (9.42) typically holds true with $\Theta = \tfrac{1}{2}$. This is certainly the case when A is coercive and A_h is a standard Galerkin approximation of A: i.e., $(A_h x_h, y_h)_Y = (A x_h, y_h)_Y$. $\quad \blacksquare$

Remark 9.4. If A is self-adjoint (or, more generally, if $A = A_1 + A_2$, with A_1 self-adjoint and $A_2: Y \supset \mathcal{D}((-A_1)^{1-\varepsilon}) \to Y$ is bounded), one can take $\Theta = \tfrac{1}{2}$ in (9.44). $\quad \blacksquare$

Theorem 9.3. (ARE) (i) The following uniform exponential stability holds true:

$$\left\| e^{(A-BB_h^*P_h)t} \right\|_{\mathcal{L}(Y)} \leq \hat{C} \, e^{-\bar{\omega}_P t}, \quad \hat{\omega}_P > 0, \tag{9.45}$$

under the same assumptions of Theorem 9.2.

(ii) Moreover,

$$\sup_{t \geq 0} e^{\hat{\omega}_P t} \left\| e^{(A-BB_h^*P_h)t} - e^{(A-BB^*P)t} \right\|_{\mathcal{L}(Y)} \to 0 \quad \text{as } h \downarrow 0. \quad \blacksquare \tag{9.46}$$

Theorem 9.2 provides the basic convergence results (with rates) for the optimal solutions of the approximating problem (9.22), (9.23), the corresponding Riccati operators, and gain operators, to the same quantities of the original problem (1.1), (1.2).

The advantage of Theorem 9.2 is this: It states that the *original* system, once acted upon by the discrete feedback control law given by $u_h^*(t,\Pi_h x) = -B^*P_h y_h^*(t,x)$ yields (uniformly) exponentially stable solutions.

Remark 9.5. Instead of the original inner product $(x_h, y_h)_Y$, one can introduce an equivalent inner product $(x_h, y_h)_{Y_h}$, where $c_1 |x_h|_Y \leq |x_h|_{Y_h} \leq c_2 |x_h|_Y$. In some situations, it is more convenient to work with a discrete inner product $(\ , \ )_{Y_h}$ as to simplify the computations for the adjoint operators for the discrete problem. $\blacksquare$

Remark 9.6. The literature on approximating schemes of optimal control problems and related Riccati equations generally assumes

(i) convergence properties of the 'open loop' solutions, i.e., of the maps $u \to y$ of the continuous problem;

(ii) "uniform stabilizability/detectability" hypotheses for the approximating problems.

In contrast, our basic assumptions are:

(a) stabilizability/detectability hypotheses (S.C.)/(D.C.) of the continuous system;

(b) a "uniform analyticity" hypothesis (A.1) on the approximations.

Starting from (a) and (b), we then derive both the convergence properties of the open loop and the uniform stabilizability/

detectability hypotheses--(i) and (ii) above--which are taken as assumptions in other treatments. Thus, the theory presented here is "optimal," in the sense that it assumes only what is strictly needed. Indeed, it can be shown that assumptions (A.1), (S.C.)/(D.C.) are not only sufficient, but also necessary, for the main theorems presented here.

These considerations are an important aspect of the entire theory since, in the case where B is an unbounded operator, the requirement of convergence $L_h \to L$ of the open loop solutions is a very strong assumption as remarked before. Generally, even when L is bounded, and the scheme is consistent, it may well happen that the scheme is not even stable; i.e., L_h may not be uniformly bounded in h. The properties of the composition $e^{At}B$ may not be retained in the approximation $e^{A_h t}B_h$. Special care must be exercised in approximating B.

Theorem 9.2 provides rate of convergence $\mathcal{O}(h^{s(1-\gamma)})$ for the approximating problem. This rate is, in general, non-optimal, as it does not reflect the regularity properties of the original continuous problem. More precisely, the regularity properties of the Riccati operator (given by (5.2)), together with the approximation property (A.2) suggests that the optimal rate of convergence of Riccati operators reconstructing this regularity should be

$$|(P_h-P)|_{\mathcal{L}(Y)} = \mathcal{O}(h^{s(1-\varepsilon)}). \tag{9.47}$$

Similarly, because of estimate (9.19), and because $\exp(A_p)t$ is analytic (Theorem 5.2), one would expect that the approximating feedback semigroup would retain convergence properties similar to (9.19), i.e.,

$$|e^{A_p t} - e^{A_{h,P_h} t}|_{\mathcal{L}(Y)} \le \frac{c\, h^{s\theta}}{t^\theta} e^{\omega t} \tag{9.48}$$

where

$$A_{h,P_h} \equiv A_h - B_h B_h^* P_h. \tag{9.49}$$

If the operator B is bounded (i.e., $B \in \mathcal{L}(U,Y)$ and $\gamma = 0$), then the above rates of convergence (9.47), (9.48) are given by Theorem 9.2. Even more, if $A^{-\gamma}B \in \mathcal{L}(U,Y)$, Theorem 9.2 provides the convergence rates equal to $\mathcal{O}(h^{s(1-\gamma)}/t^{1-\varepsilon})$, which are 'nonoptimal' if $\gamma > 0$. Thus, the following question arises: is it possible to obtain the optimal rates at convergence, (9.47) and (9.48) in the unbounded case, i.e., when $A^{-\gamma}B \in \mathcal{L}(U;Y)$, $\gamma > 0$ (particularly, in the interesting case $\gamma > \frac{1}{2}$)? Below we shall provide a positive answer to the above question, provided, however, very special care is given to the selection of the approximations A_h, B_h. While the convergence results with the rate $\mathcal{O}(h^{s(1-\gamma)}/t^{1-\varepsilon})$ are valid for any consistent approximations A_h, B_h (for instance, $B_h = B$ or $B_h = \Pi_h B$) subject to (A.1)-(A.6), the optimal rates of convergence (i.e., (9.47), (9.48)) will require, in general, additional hypotheses imposed on the approximations of the unbounded operator B.

Finally, it should be noted that in the case of B-unbounded, the optimal rates require a more delicate analysis. This is so, because of the necessity of 'tracing' the singular behavior (at the origin) of the optimal solutions. The crucial role to this end is played by the so-called 'rough data' estimates together with a perturbation result which asserts, roughly speaking, that relatively bounded perturbations preserve 'uniform analyticity' and 'uniform stabilizability' with estimates independent of the parameter of discretization.

Below, we shall formulate our main abstract results.

The additional approximation properties for the operators B_h and A_h are the following.

(A.7)   Let $U_{r_0} \subseteq U \subseteq U_{r_1}$ be two Hilbert spaces such that

   (i) $|[\hat{A}_h^{-1}B_h - \hat{A}^{-1}B]u|_Y \leq C h^\rho |u|_{U_{r_1}}$;

   (ii) $|[\hat{A}_h^{-1}B_h - \hat{A}^{-1}B]u|_Y \leq C h^{r_0} |u|_{U_{r_0}}$,

   where $0 \leq r_0 \leq s$, and $\rho > 0$.

(A.8)   Let Y_{r_1} be another Hilbert space such that for some $\varepsilon > 0$,

   $Y_{r_1} \supset \mathcal{D}(\hat{A}^{1-\varepsilon}) \cap \mathcal{D}(\hat{A}^{*1-\varepsilon})$ and

(i) $\hat{A}^{-1+\varepsilon}B \in \mathcal{L}(U_{r_1};Y)$; $\hat{A}^{-1}B \in \mathcal{L}(U_{r_0};Y_{r_1})$,

(ii) $|[B_h^*\hat{A}_h^{*-1}-B^*\hat{A}^{*-1}]y|_{U_{r_1}} \leq C h^{r_1}|y|_{Y_{r_1}}$,

where $0 \leq r_1 \leq s$.

(A.9) (i) $B^*\hat{A}^{*-2+\varepsilon} \in \mathcal{L}(Y;U_{r_0})$; $B^*\hat{A}^{*-1+\varepsilon}\hat{A}^{-1+\varepsilon} \in \mathcal{L}(Y;U)$.

(ii) There exists $n \geq 1$ such that

$$[B^*\hat{A}^{*-1}\hat{A}^{-1}B]^n \in \mathcal{L}(U,U_{r_0}).$$

Theorem 9.4. [Las.6] In addition to the hypotheses of Theorem 9.2, assume hypotheses (A.7)-(A.9). Then with $\varepsilon > 0$ arbitrarily small, and C independent of h and t,

(i) $|P-P_h|_{\mathcal{L}(Y)} \leq C[h^{s(1-\varepsilon)}+h^{r_0}+h^{r_1}]$;

(ii) $|B_h^*P_h-B^*P|_{\mathcal{L}(Y)} \leq C h^{-\gamma(s+\varepsilon)}[h^s+h^{r_0}+h^{r_1}]$.

Theorem 9.5. [Las.6] Assume the same hypotheses as above. Then, there exists $\omega_0 > 0$ such that for any $\varepsilon > 0$, $t \geq 0$,

(i) $|(y^0-y_h^0)(t)|_Y \leq \dfrac{Ce^{-\omega_0 t}}{t^{1-\varepsilon}} |x|_Y[h^{s(1-\varepsilon)}+h^{r_0}+h^{r_1}]$;

(ii) $|(u^0-u_h^0)(t)|_U \leq \dfrac{Ce^{-\omega_0 t}}{t^{\gamma-\varepsilon}} |x|_Y[h^{s(1-\varepsilon)}+h^{r_0}+h^{r_1}]$;

(iii) $|e^{A_P t}-e^{A_{P_h} t}|_{\mathcal{L}(Y)}+|e^{A_P t}-e^{(A-BB^*P_h)t}|_{\mathcal{L}(Y)}$

$$\leq C e^{-\omega_0 t}h^{-\gamma(s+\varepsilon)}[h^s+h^{r_0}+h^{r_1}];$$

(iv) $|e^{A_P t}-e^{A_{P_h} t}|_{\mathcal{L}(Y)} \leq \dfrac{Ce^{-\omega_0 t}}{t^{\gamma-\varepsilon}} [h^{s(1-\varepsilon)}+h^{r_0}+h^{r_1}]$.

Corollary 9.6. Let $x \in \mathcal{D}(A)$, then

$$|(u^0-u_h^0)(t)|_U \leq C e^{-\omega_0 t}[h^{s(1-\varepsilon)}+h^{r_0}+h^{r_1}]|x|_Y;$$

$$|(y^0 - y_h^0)(t)|_Y \leq C e^{-\omega_0 t} [h^{s(1-\epsilon)} + h^{r_0} + h^{r_1}] |x|_Y.$$

Corollary 9.6 follows easily from Theorem 9.5 and the usual "boost-strap" argument as in [F.1] or [L-T.2], [L-T.19].

Theorem 9.5 gives the optimal rates of convergence provided $r_0 = r_1 = s$. On the other hand, the values of r_0 and r_1 depend on the 'goodness' of the approximations B_h. If the operator B is bounded, i.e., $B \in \mathcal{L}(U;Y)$, then one can take $B_h = \Pi_h B$ or $B_h = B$, and the additional assumptions (A.7)-(A.9) are automatically satisfied with $r_0 = r_1 = s$. Indeed, it is enough to take $U_{r_0} = U$ and (A.7) (with $r_0 = s$) follows from the hypothesis (A.2). Similarly, in (A.8) we take $Y_{r_1} = Y$, $U_{r_1} = U$, and $r_1 = s$. (A.9) holds trivially if $B \in \mathcal{L}(U;Y)$.

In the more general case where the operator B is unbounded and one takes $B_h = \Pi_h B$ or $B_h = B$, then assumptions (A.7), (A.9) hold true with $r_0 = r_1 = s(1-\gamma)$. In fact, we have

Corollary 9.7. Assume (A.1)-(A.6). Let $B_h = \Pi_h B$. Then the statements of Theorems 9.4 and 9.5 hold true with $r_0 = r_1 = s(1-\gamma)$. ∎

In view of the results of Theorems 9.4 and 9.5, in order to obtain the optimal rates of convergence in the case of unbounded operator B, it is necessary that the approximations A_h and B_h comply with the additional hypotheses (A.6)-(A.9) where $r_0 = r_1 = s$. We shall show in Section 10 in concrete examples of the heat equation with boundary controls that such approximations can indeed be constructed. For instance, in the case of parabolic problems with Dirichlet boundary controls, "Nitsche scheme" provides an example of an approximating algorithm satisfying all the requirements (A.6)-(A.9). In the Neumann case, the usual Galerkin approximation of the elliptic operator complies with assumptions (A.6)-(A.9). Similar examples can be provided for the strongly damped plate equations with either boundary or point controls (see Section 10).

The approximation Theorems 9.2 and 9.3 in the general case $\gamma < 1$ is proved in [L-T.19]. Here the techniques rely on a combination of ideas of the continuous problem [L-T.8], [F.1], [D-I], together with approximating properties of analytic semigroups [L.1] and convergence properties for the open loop problem, [Las.2]. The proof of Theorem 9.1 on the DRE is simpler, and it follows from the arguments given in [L-T.19]. Theorems 9.4 and 9.6 are proved in [Las.6].

9.2. Approximation for the (H.2)-class

9.2.1 Approximating assumptions

Approximation of A. We assume as $h \downarrow 0$:

(B.1)
$$\hat{A}_h^{-1}\Pi_h - \Pi_h\hat{A}^{-1} \to 0 \quad \text{strongly in Y;}$$

$$\hat{A}_h^{*-1}\Pi_h - \Pi_h\hat{A}^{*-1} \to 0 \quad \text{in Y.} \qquad (9.50)$$

(B.2)
$$|e^{A_h t}|_{\mathcal{L}(Y)} \le C\, e^{\omega t}, \quad t > 0. \qquad (9.51)$$

Approximation of B. We assume as $h \downarrow 0$:

(B.3)
$$\int_0^T |B_h^* e^{A_h^* t}\Pi_h x|_U^2\, dt \le C_T |x|_Y^2 \quad \text{(discrete analog of (H.2)).} \qquad (9.52)$$

With reference to L and L_h defined by (9.10) and (9.11), we assume

(B.4)
$$|(L_h-L)u|_{C([0,T];Y)} \to 0 \quad \text{for } u \in L_2(0,T;U); \qquad (9.53a)$$

$$|(L_h^*\Pi_h-L^*)f|_{L_2(0,T;U)} \to 0 \quad \text{for } f \in L_1(0,T;Y). \qquad (9.53b)$$

Sufficient conditions for assumptions (B.4) to hold are the following assumptions [Las.3]: (B.1)-(B.3) together with

(B.4$_s$)

(i)
$$|\hat{A}^{-1}(B_h-B)u|_Y \to 0; \quad u \in U; \qquad (9.54)$$

(ii)
$$|(\hat{A}_h^{-1}-\hat{A}^{-1})B_h u|_Y \to 0; \quad u \in U; \qquad (9.55)$$

(iii) $$|(B_h^* \Pi_h - B^*)\hat{A}^{*-1}x|_U \to 0; \quad x \in Y; \qquad (9.56)$$

(iv) $$|B_h^*(\hat{A}_h^{*-1}\Pi_h - \Pi_h\hat{A}^{*-1})x|_U \to 0; \quad x \in Y. \qquad (9.57)$$

9.2.2. Approximation of dynamics and of control problem. Related Riccati Equation

Control problem. Given the approximating dynamics $y_h(t) \in V_h$ such that

$$\dot{y}_h(t) = A_h y_h(t) + B_h u(t), \quad y_h(0) = \Pi_h y(0) \qquad (9.58)$$

minimize

$$J(u, y_h(u)) = \int_0^T [|Ry_h(t)|_Z^2 + |u(t)|_U^2]dt \qquad (9.59)$$

with $T < \infty$ for the DRE and $T = \infty$ for the ARE.

Riccati Equation. The approximating Riccati Equations are given in $(DRE)_h = (9.3)$ and $(ARE)_h = (9.4)$.

9.2.3. Approximating results

Differential Riccati Equations

Theorem 9.8. (DRE) [Las.5] I. Assume hypothesis (H.2) = (1.6) and (5.0) for the continuous problem and the approximation hypotheses (B.1) = (9.50) through (B.4) = (9.53). Then, as $h \downarrow 0$:

(i) $$|P_h(\cdot)\Pi_h x - P(\cdot)x|_{C([0,T];Y)} \to 0, \quad x \in Y; \qquad (9.60)$$

(ii) $$|y_h^0(\cdot, \Pi_h x) - y^0(\cdot, x)|_{C([0,T];Y)} \to 0, \quad x \in Y; \qquad (9.61)$$

(iii) $$|J(u_h^0, y_h^0) - J(u^0, y^0)| \to 0. \qquad (9.62)$$

II. In addition, assume that as $h \downarrow 0$,

$$|\int_0^T B^*[e^{A_h^*t}\Pi_h - e^{A^*t}]R^*Rg(t)dt|_U \to 0, \quad g \in C([0,T];Y). \qquad (9.63)$$

Then as $h \downarrow 0$:

$$|B^*[P_h(\cdot)\Pi_h - P(\cdot)]x|_{C([0,T];U)} \to 0, \quad x \in Y. \quad \blacksquare \quad (9.64)$$

Remark 9.7. Note that in Part I of Theorem 9.8, in order to obtain convergence to the optimal solutions and to the Riccati operator in (9.60)-(9.62), no smoothing assumption on R is imposed; here simply $R \in \mathcal{L}(Y,Z)$ as in (5.0). However, in Part II, in order to obtain convergence of the gain operators $B^*P_h(t)$, it is essential that R^*R has a regularizing effect as postulated by (9.63). It can be easily shown--for example when exp(At) is a group--that conclusion (9.64) in general fails with $R \in \mathcal{L}(Y;Z)$ only. See comments below Corollary 5.4. $\blacksquare$

Theorem 9.8 is proved in [Las.5]: It is enough to notice that assumptions (B.1)-(B.2) imply hypothesis (3.7) of Theorem 3.1 in [Las.5].

Algebraic Riccati Equation. In the more delicate approximation case of ARE, we need a discrete counterpart of the Finite Cost Condition (1.9), which would then guarantee solvability of the finite dimensional $ARE_h = (9.4)$. Also, in contrast with the results of Theorem 9.8, Part II, on the DRE, no smoothing assumption on the observation R is needed. In this case, one obtains convergence of the gain operators, but only as unbounded operators (9.75) below. This, again, is in line with the continuous theory.

Theorem 9.9. [Las.3]. Assume
    I. the continuous hypotheses (H.2), (F.C.C.) = (1.9) and (D.C.) = (5.17)-(5.19).
    II. The approximation properties (B.1)-(B.4) and, in addition:
$(F.C.C.)_h$ (uniform Finite Cost Condition):

$$\exists \, \alpha > 0; \, \forall \, y_0 \in Y, \, \exists \, u_h \in L_2(0,\infty;U)$$

such that $J(u,y_h(u)) \leq \alpha|y_0|_Y^2.$ \quad (9.65)

$(D.C.)_h$ (uniform Detectability Condition): There exist $K_h: Z \to V_h$ such that

$$|K_h^*x_h|_Z \leq C[|B_h^*x_h|_U + |x_h|_Y], \quad (9.66)$$

and

$$|e^{A_{K_h}t}|_{\mathcal{L}(Y)} \le C e^{-\omega_1 t}, \tag{9.67}$$

where $A_{K_h} = A_h - K_h R$.   Then:

   I.   (convergence of Riccati operators)

$$|P_h \Pi_h x - Px|_Y \to 0, \quad x \in Y, \tag{9.68}$$

$$|e^{A_{P,h}t} x_n|_Y \le C e^{-\bar{\omega}_0 t} |x_h|_{Y'} \tag{9.69}$$

where $A_{P,h} = A_h - B_h B_h^* P_h$.

   II.   (convergence of optimal solutions)

$$|u_h^0 - u^0|_{L_2(0,\infty;U)} \to 0; \tag{9.70}$$

$$|y_h^0 - y^0|_{L_2(0,\infty;Y)} \to 0; \tag{9.71}$$

$$|y_h^0 - y^0|_{C(0,\infty;Y)} \to 0. \tag{9.72}$$

Hence

$$|e^{A_{P,h}t} \Pi_h x - e^{A_P t} x|_{L_2(0,\infty;Y)} \to 0; \tag{9.73}$$

$$|e^{A_{P,h}t} \Pi_h x - e^{A_P t} x|_{C(0,\infty;Y)} \to 0. \tag{9.74}$$

   III.   (convergence of "gain" operators)

(i) $$|B_h^* P_h \Pi_h x|_U \to |B^* Px|_{U'} \quad x \in \mathcal{D}(A). \tag{9.75}$$

(ii)   For each $x \in \mathcal{D}(A_F)$ there exists a sequence $x_h \in Y_h$ such that $x_h \to x$ in Y and

$$|B_h^* P_h x_h - B^* Px| \to 0. \quad \blacksquare \tag{9.76}$$

   The proof of the above theorem is given in [Las.3]. This theorem provides us with the convergence theory for the Riccati operators and the gain feedbacks with minimal assumptions imposed on the model. We notice that the convergence of the gain operator holds on a dense set in Y, and not on the whole space Y. This is consistent with the continuous theory, where the gain operator $B^* P$ is only densely defined.

Remark 9.8. Note that in both cases (general abstract C_0 semigroup and analytic semigroup) the above results are optimal, as they reconstruct numerically the properties of the solution which are present in the continuous case. This means that in the case of general C_0 semigroups and unbounded operators B, we obtain strong convergence of the Riccati operators on the full space Y, while the gain operators converge as unbounded operators on some dense set. In the analytic case, the uniform convergence of both Riccati operators and gain operators hold on the entire space Y. This, again, is in agreement with the regularity of the continuous theory (i.e., our approximation reconstructs "optimally" the properties of the continuous problem). ∎

9.2.4. **Discussion on the assumptions**

(i)   Note that in Theorem 9.9--in contrast with the analytic situation of Section 9.1--we need to assume the "uniform Finite Cost Condition" $(F.C.C.)_h = (9.65)$ and the "uniform Detectability Condition" $(D.C.)_h = (9.66)$, (9.67). In the analytic case subject to assumption $(H.1) = (1.5)$, these properties can be deduced from the "uniform analyticity condition" $(A.1) = (9.13)$. Instead, in the case of an arbitrary s.c. semigroup, these properties need to be established independently for a specific choice of the approximation scheme. Indeed, these conditions are in general rather sensitive and scheme dependent. They may fail, even in the case of B bounded, if an inappropriate approximating scheme is selected. Negative examples are known, even in the case of retarded (delay) equations, with spline approximations [L-M], [M.1], [P]. These conditions are related to the following fact: The spectrum of the original operator A should be "faithfully" approximated by the chosen approximating scheme.

(ii)   All the remaining assumptions (B.1)-(B.4) are very natural and, in fact, minimal ones. They are consistent with the hypotheses imposed on the continuous problem. Indeed, (B.1) and (B.2) are the usual requirements of consistency of the approximation of the original semigroup and its adjoint.

Hypothesis (B.3) is a discrete counterpart of (H.2), while the assumptions grouped in $(B.4_s)$ are in line with the continuous property that $\hat{A}^{-1}B$ be bounded (and, in fact, they are satisfied if one takes $B_h = \Pi_h B$).

9.2.5. <u>Literature</u>

Most of the literature dealing with approximation schemes for Riccati Equations for arbitrary C_0-semigroup treats the case of the input operator B bounded, see e.g., [G], [I.1], [KS]. In the B-unbounded case and with arbitrary C_0-semigroups, we are aware, in addition to [Las.3], [Las.5] of only one paper [I-T] where the approximations of ARE are discussed subject to the condition (H.2) and the additional requirement that the observation R is smoothing like in [P-S] of Part I. Since, as already discussed, the framework of [P-S] is not applicable to all the examples of Sections 9.1, 9.2, 9.3, 9.4 the treatment of [I-T] cannot be applied either.

10. <u>Examples of numerical approximation for the classes (H.1) and (H.2)</u>

Except for the case of first-order hyperbolic systems, in this section we shall concentrate only on the more demanding approximation case for the ARE, where more conditions need to be satisfied. We shall illustrate the applicability of the approximation Theorem 9.3-9.6 (class (H.1)) and of the approximation Theorem 9.9 (class (H.2)) in a few examples taken from the continuous Sections 6, 7. For a full treatment of the case of the heat equation with Dirichlet boundary control, we refer to [L-T.1].

10.1. <u>Class (H.1): Heat equation with Dirichlet boundary control</u>

We return to the continuous problem of section 6.1, to which we apply the approximating theory [L-T.1], [L-T.19], [Las.6].

<u>Choice of</u> V_h. We shall select as the approximating space $V_h \subset H_0^1(\Omega)$ to be a space of splines (linear, quadratic, etc.) which comply with the usual approximation properties:

$$\|\Pi_h y - y\|_{H^\ell(\Omega)} \leq C\, h^{s-\ell} \|y\|_{H^s(\Omega)}, \quad s \leq 2;\ s-\ell \geq 0;\ 0 \leq \ell \leq 1; \quad (10.1)$$

inverse approximation properties [B]:

$$\|y_h\|_{H^\alpha(\Omega)} \leq C\, h^{-\alpha} \|y_h\|_{L_2(\Omega)}, \quad 0 \leq \alpha \leq 1, \quad (10.2i)$$

$$\left\|\frac{\partial}{\partial \nu} (y - \Pi_h y)\right\|_{L_2(\Gamma)} \leq C h^{s-\frac{1}{2}} \|y\|_{H^s(\Omega)}, \quad \frac{1}{2} < s \leq 2, \tag{10.2ii}$$

$$\|y_h\|_{L_2(\Gamma)} + h\left\|\frac{\partial y_h}{\partial \nu}\right\|_{L_2(\Gamma)} \leq C h^{-\frac{1}{2}} \|y_h\|_{L_2(\Omega)}, \quad y_h \in V_h, \tag{10.2iii}$$

where Π_h is the orthogonal projection of $L_2(\Omega)$ onto V_h.

Choice of A_h. We define $A_h: V_h \to V_h$ as usual, where the inner products are in L_2:

$$(A_h x_h, y_h)_\Omega = (A x_h, y_h)_\Omega$$

$$= -\int_\Omega \nabla x_h \cdot \nabla y_h \, d\Omega + c^2 (x_h, y_h)_\Omega, \quad x_h, y_h \in V_h. \tag{10.3}$$

Choice of B_h. With reference to (6.5), we define $B_h: U \to V_h$ by

$$B_h = -Q_h A D_1, \tag{10.4a}$$

D_1 as in (6.6), (6.7), and we notice that (L_2-inner products)

$$(B_h u, y_h)_\Omega = -(A D_1 u, y_h)_\Omega = -(u, D_1^* A y_h)_\Gamma = (u, \frac{\partial y_h}{\partial \nu})_\Gamma. \tag{10.4b}$$

Hence

$$B_h^* y_h = \frac{\partial y_h}{\partial \nu}. \tag{10.5}$$

Approximating control problem. This is given by the O.D.E. problem:

$$\begin{cases} (\dot{y}_h, \phi_h)_\Omega + \int_\Omega \nabla y_h \cdot \ell \phi_h \, d\Omega - c^2 \|y_h\|_\Omega^2 = (u, \frac{\partial}{\partial \nu} \phi_h)_\Gamma, \quad \phi_h \in V_h; \\ (y_h(0), \phi_h)_\Omega = (y(0), \phi_h)_\Omega. \end{cases} \tag{10.6}$$

The optimal feedback control for the approximating finite-dimensional problem is

$$u_h^0(t; 0, y_0) = -\frac{\partial}{\partial \nu} P_h y_h^0(t; 0, y_0),$$

where P_h satisfies the following discrete Algebraic Riccati Equation

$$-\int_\Omega \nabla P_h x_h \cdot \nabla y_h d\Omega - \int_\Omega \nabla x_h \cdot \nabla P_h y_h d\Omega + (x_h, y_h)_\Omega = (\frac{\partial}{\partial\nu} P_h x_h, \frac{\partial}{\partial\nu} P_h y_h)_\Gamma ,$$

$$\forall \, x_h, y_h \in V_h . \qquad (10.7)$$

Verification of assumptions of Theorem 9.2. **Assumptions (9.31) and (9.32)**. These are plainly satisfied since $R = I$ and $\hat{A}^{-\gamma} B \in \mathcal{L}(U, Y)$ with $\gamma = \frac{3}{4} + \varepsilon$ in our case. Because of the compactness of A^{-1} (since Ω is bounded), this then implies in turn that $\hat{A}^{-1} B$ is compact $U \to Y$, and thus $B^*(\hat{A}^*)^{-1}$ is compact $Y \to U$, as desired.

Assumption (A.1) = (9.13) (uniform analyticity). That this is satisfied follows from results on Galerkin approximations of elliptic operators, see [B-S] for the self-adjoint case and [Las.1] for the general non-self-adjoint case.

Assumption (A.2) = (9.14). The standard elliptic approximation is

$$\|\Pi_h \hat{A}^{-1} - \hat{A}_h^{-1} \Pi_h\|_{\mathcal{L}(L_2(\Omega))} \leq C \, h^2$$

[B-A], so that (A.2) holds with $s = 2$.

Assumption (A.3) = (9.15). By (10.5) and (10.2iii), we obtain with $U = L_2(\Gamma)$ and $Y = L_2(\Omega)$,

$$\|B^* y_h\|_U = \|B_h^* y_h\|_U = \|\frac{\partial}{\partial\nu} y_h\|_{L_2(\Gamma)} \leq C \, h^{-\frac{1}{2}} \|y_h\|_{L_2(\Omega)} . \qquad (10.8)$$

Thus (A.3) is satisfied (conservatively) for $s = 2$, $\gamma = \frac{3}{4} + \varepsilon$.

Assumption (A.4) = (9.16). By (10.5) and (10.2ii) applied with $s = 2$,

$$\|B^*(\Pi_h x - x)\|_{L_2(\Gamma)} = \|\frac{\partial}{\partial\nu}(\Pi_h x - x)\|_{L_2(\Gamma)} \leq C \, h^{\frac{1}{2}} \|x\|_{H^2(\Omega)} , \qquad (10.9)$$

which implies (A.4) in view of the fact that $\mathcal{D}(A) \subset H^2(\Omega)$ and $s(1-\gamma) = 2(1 - \frac{3}{4} - \varepsilon) = \frac{1}{2} - 2\varepsilon < \frac{1}{2}$.

Assumption (A.5) = (9.17). Since in our case $B_h^* \Pi_h = B^* \Pi_h$, (A.4) coincides with (A.5).

Assumption (A.6) = (9.18). From (10.2iii) applied with $s = \frac{1}{2}+\varepsilon$ and from the trace theorem, we obtain

$$\|B^*\Pi_h x\|_{L_2(\Gamma)} = \|\frac{\partial}{\partial\nu}\Pi_h x\|_{L_2(\Gamma)} \leq \|\frac{\partial}{\partial\nu}(\Pi_h-I)x\|_{L_2(\Gamma)} + \|\frac{\partial}{\partial\nu}x\|_{L_2(\Gamma)}$$

$$\leq C h^\varepsilon \|x\|_{H^{\frac{1}{2}+\varepsilon}(\Omega)} + C\|x\|_{H^{\frac{1}{2}+\varepsilon}(\Omega)} . \qquad (10.10)$$

(A.6) follows now from $\mathcal{D}(A^{*\frac{1}{4}+\varepsilon}) \subset H^{\frac{1}{2}+2\varepsilon}(\Omega)$.

Thus, we have verified all the assumptions of Theorems 9.2 and 9.3 in the case of the heat equation problem with Dirichlet boundary control as in (10.1). Then, application of Theorem 9.2 yields the following convergence results:

(i) $\qquad \|P_h\Pi_h-P\|_{\mathcal{L}(L_2(\Omega))} \leq C h^{\varepsilon_0}, \quad \varepsilon_0 < \frac{1}{2};$ $\qquad\qquad\qquad (10.11)$

(ii) $\qquad \|\frac{\partial}{\partial\nu}[P_h\Pi_h-P]\|_{\mathcal{L}(L_2(\Omega);L_2(\Gamma))} \to 0 \quad$ as $h\downarrow 0;$ $\qquad\quad (10.12)$

(iii) $\quad \|y_h^0-y^0\|_{\mathcal{L}(L_2(\Omega);L_2(0,\infty;L_2(\Omega)))}$

$$+ \sup_{t\geq 0} e^{\overline{\omega}_p t} t^\varepsilon \|y_h^0(t)-y^0(t)\|_{\mathcal{L}(L_2(\Omega))} \leq C h^{\varepsilon_0}, \quad \varepsilon_0 < \frac{1}{2}. \quad (10.13)$$

Moreover, if we use the feedback law given by

$$\hat{u}_h(t) = -\frac{\partial}{\partial\nu} P_h y(t),$$

and we insert it into the original dynamics

$$\begin{cases} y_t = (\Delta+c^2)y \\ y_{|\Sigma} = \hat{u}_h \end{cases}$$

then the corresponding system is exponentially stable in $L_2(\Omega)$ uniformly in the parameter h.

Remark 10.1. The rate of convergence $\mathcal{O}(h^{\frac{1}{2}-\varepsilon})$ guaranteed by (10.11)-(10.13) is not optimal. In view of the regularity $P \in \mathcal{L}(L_2(\Omega);H^{2-\varepsilon}(\Omega))$ of the Riccati operator (see (5.2)), one would expect that the optimal rate of convergence should be of the order of $\mathcal{O}(h^{2(1-\varepsilon)})$. Indeed, we shall show that this is possible, but for different, appropriate approximations of A_h and B_h. More precisely, in order to obtain the optimal rates of convergence $(\mathcal{O}(h^{2(1-\varepsilon)}))$, care must be exercised in selecting the approximation of the Poisson operator A^{-1}. Since the Dirichlet problem does not admit a natural variational formulation, extra attention must be paid to the approximation of the boundary conditions. Thus, in order to obtain the optimal rate $(\mathcal{O}(h^{2(1-\varepsilon)}))$, we need to introduce an approximation which approximates 'well' the boundary conditions. For this purpose we shall use the elliptic approximation of the Poisson operator due to Nitsche [N.1].

With V_h defined by (10.1), (10.2) with $s > \frac{1}{2}$, let $A_h \colon V_h \to V_h$ be defined as (see [N.1])

$$(A_h x_h, y_h) \equiv \tilde{a}(x_h, y_h) \equiv a(x_h, y_h) - (\tfrac{\partial}{\partial \nu} x_h, y_h)_\Gamma$$

$$- (x_h, \tfrac{\partial}{\partial \nu} y_h)_\Gamma + \beta\, h^{-1}(x_h, y_h)_\Gamma + c^2(x_h, y_h)_\Omega \qquad (10.14)$$

in the L_2-norms, where $\beta > 0$ is sufficiently large and c^2 as in (6.1a).

The approximating finite-dimensional Riccati operator $P_h \colon V_h \to V_h$ satisfies the following Approximating Algebraic Riccati Equation:

(ARE$_h$) $\qquad -(A_h P_h x_h, y_h) - (P_h A_h x_h, y_h) + (x_h, y_h)$

$$= ((\tfrac{\partial}{\partial \nu} - \beta h^{-1}) P_h x_h, (\tfrac{\partial}{\partial \nu} - \beta h^{-1}) P_h y_h)_\Gamma. \qquad (10.15)$$

We shall now verify the assumptions of Theorems 9.4 and 9.5 on optimal rates.

Hypotheses (A.1) = (9.13) - (A.2) = (9.14) (with s = 2) are well known for the Nitsche's approximation A_h defined in (10.14) (see [B-S-T-W]).

Hypotheses (A.3) = (9.15) - (A.5) = (9.17). In the Dirichlet case we have (see [Ch-L])

$$B_h^* x_h = \tfrac{\partial}{\partial \nu} x_h + \beta h^{-1} x_h \big|_\Gamma . \qquad (10.16)$$

Thus, hypothesis (A.3) is the result of the inverse approximation property (10.2iii). Since $B^* x = \tfrac{\partial}{\partial \nu} x$ and for $x \in \mathcal{D}(A)$, we have $x|_\Gamma = 0$, we obtain

$$|(B^*-B_h^*)x|_U = |\frac{\partial}{\partial\nu} x + \beta h^{-1}x|_\Gamma - \frac{\partial}{\partial\nu} \Pi_h x - \beta h^{-1}\Pi_h x|_{L_2(\Gamma)}$$

$$\leq |\frac{\partial}{\partial\nu}(\Pi_h-I)x|_{L_2(\Gamma)} + \beta h^{-1}|(\Pi_h-I)x|_{L_2(\Gamma)}$$

(by the approximation property (10.2ii) and (10.1))

$$\leq C h^{-\frac{1}{2}}h^2|x|_{H^2(\Omega)} \leq C h^{2(1-\frac{1}{4})}|x|_{\mathcal{D}(A)} \qquad (10.17)$$

as desired for (A.5) to hold. As for (A.4), we have as desired

$$|B^*(I-\Pi_h)x|_U = |\frac{\partial}{\partial\nu}(I-\Pi_h)x|_U \leq C h^{\frac{1}{2}}|x|_{H^2(\Omega)}. \qquad (10.18)$$

Hypothesis (A.6) = (9.18). It involves only B^* (not A_h, B_h^*) and was verified before in (10.10).

Hypothesis (A.7). Let $z_h = \hat{A}_h^{-1}B_h u$ and $z = \hat{A}^{-1}Bu$. We have

$$\tilde{a}(z_h,x_h)+\omega(z_h,x_h) = (u, \frac{\partial}{\partial\nu} x_h)_\Gamma + Bh^{-1}(u,x_h)_\Gamma , \qquad (10.19)$$

and

$$A(\xi,\partial)z+\omega z = 0, \quad z|_\Gamma = u. \qquad (10.20)$$

Since (10.19) defines an elliptic approximation of z, the convergence results of [N.1] apply to yield

$$|z-z_h|_{L_2(\Gamma)} \leq C h^2|z|_{H^2(\Omega)} \leq C h^2|u|_{H^{\frac{1}{2}}(\Gamma)} , \qquad (10.21)$$

so (A.7)(ii) holds with $r_0 = 2$ and $U_{r_0} = H^{\frac{1}{2}}(\Gamma) \subset L_2(\Gamma)$.

As for (A.7)(i), we shall prove that

$$|[\hat{A}_h^{-1}B_h-\hat{A}^{-1}B]|_{\mathcal{L}(L_2(\Gamma);L_2(\Omega))} = \mathcal{O}(h^{\frac{1}{2}}), \qquad (10.22)$$

so (A.7)(i) is satisfied with $\rho = \frac{1}{2}$ and $U_{r_1} = L_2(\Gamma)$.

To assert (10.22), we use a duality argument,

$$|[B^*\hat{A}^{*-1}-B_h^*\hat{A}_h^{*-1}]x|_U = |\frac{\partial}{\partial\nu}(\hat{A}_h^{-1}-\hat{A}^{-1})x|_{L_2(\Gamma)}$$

$$+ \beta h^{-1}|(\hat{A}_h^{-1}-\hat{A}^{-1})x|_{L_2(\Gamma)} \leq C h^{-\frac{1}{2}}h^2|x|_{L_2(\Omega)}.$$

Hypothesis (A.8). Here we have $U_{r_1} = U = L_2(\Gamma)$ and $H_{r_1} = H^{\frac{3}{2}}(\Omega) \supset \mathcal{D}(A^{\frac{3}{4}})$; $r_1 = 2$. Part (i) of (A.8) is trivially satisfied. As for part (ii), we compute

$$|[B_h^* \hat{A}_h^{*-1} - B^* \hat{A}^{*-1}]y|_Y \leq |B_h^*[A_h^{*-1} - \Pi_h \hat{A}^{*-1}]y|_{L_2(\Gamma)} + |[B_h^* \Pi_h \hat{A}^{*-1} - B^* \hat{A}^{*-1}]y|_{L_2(\Gamma)}$$

(by (A.3)-(A.5)) $\qquad \leq h^{-\frac{3}{2}} h^r |A^{-1}y|_{H^r(\Omega)} + h^{-\frac{3}{2}} h^r |\hat{A}^{-1}y|_{H^r(\Omega)}$

if r in (10.1) $\geq 3\frac{1}{2}$
$$\leq C h^{r-\frac{3}{2}} |y|_{H^{r-2}(\Omega)} \leq C h^2 |y|_{H^{\frac{3}{2}}(\Omega)}. \tag{10.23}$$

Hypothesis (A.9). We have

$$B^* \hat{A}^{*-1} \hat{A}^{-1} B = \frac{\partial}{\partial \nu} \hat{A}^{-2} AD = \frac{\partial}{\partial \nu} \hat{A}^{-1} \hat{A}^{-1} AD. \tag{10.24}$$

From elliptic theory [L-M],

$$\hat{A}^{-1} AD \in \mathcal{L}(H^\alpha(\Gamma); H^{\alpha+\frac{1}{2}}(\Omega)), \quad \alpha \geq 0, \tag{10.25}$$

$$\hat{A}^{-1} \hat{A}^{-1} AD \in \mathcal{L}(H^\alpha(\Gamma); H^{\alpha+\frac{5}{2}}(\Omega)). \tag{10.26}$$

Thus, applying (10.26) with $\alpha = \frac{1}{2}$ and trace theory yields

$$B^* \hat{A}^{*-1} \hat{A}^{-1} B \in \mathcal{L}(H^{\frac{3}{2}}(\Gamma); H^{\frac{5}{2}}(\Gamma)), \tag{10.27}$$

which proves part (i) of (A.9) (we recall that $U_{r_0} = H^{\frac{3}{2}}(\Gamma)$).

For part (ii) we use

$$B^* \hat{A}^{*-2+\varepsilon} = \frac{\partial}{\partial \nu} \hat{A}^{-2+\varepsilon},$$

and since

$$\hat{A}^{-2+\varepsilon} \in \mathcal{L}(L_2(\Omega); H^{4-2\varepsilon}(\Omega)). \tag{10.28}$$

Trace Theorem implies part (ii) of (A.9).

Finally, part (iii) is satisfied with $n = 2$. Indeed, (10.26), applied with $\alpha = 0$ and trace theory, gives

$$B^* \hat{A}^{*-1} \hat{A}^{-1} B \in \mathcal{L}(L_2(\Gamma); H^1(\Gamma)). \tag{10.29}$$

Repeated application of (10.26) this time with $\alpha = 1$ gives

$$B^*\hat{A}^{*-1}\hat{A}^{-1}B \in \mathcal{L}(H^1(\Gamma); H^2(\Gamma)). \qquad (10.30)$$

Combining (10.29) with (10.30) gives the desired result of (A.9)(iii).

Thus, we have verified all assumptions of Theorems 9.4 and 9.5. Application of these theorems to the heat equation (3.1) with Dirichlet control yields the following results.

Theorem 10.1. Assume that $\frac{1}{2} < s \leq 2$ in (10.1). Then

I. The unique solution $P_h: V_h \rightarrow V_h$ to $(ARE_h) = (10.15)$ satisfies the following estimate:

$$|P_h - P|_{\mathcal{L}(L_2(\Omega))} \leq C\, h^{2(1-\varepsilon)}.$$

II. There exists $\omega_0 > 0$ and $C > 0$ such that

$$|y_h^0(t)|_{L_2(\Omega)} \leq C\, e^{-\omega_0 t}\, |y_0|_{L_2(\Omega)},$$

where $y_h^0(t)$ satisfies in the L_2-norms

$$\begin{cases} (\dot{y}_h^0(t), x_h) + \tilde{a}(y_h^0(t), x_h) = ((\frac{\partial}{\partial\nu} - \beta h^{-1})P_h y_h^0(t), x_h)_\Gamma \\ (y_h^0(0), x_h) = (y_0, x_h). \end{cases}$$

III.
$$|y_h^0(t) - y^0(t)|_{L_2(\Omega)} \leq \frac{C\, h^{2(1-\varepsilon)}}{t^{1-\varepsilon}}\, e^{-\omega_0 t}\, |y_0|_{L_2(\Omega)}.$$

IV.
$$|y_h^*(t) - y^0(t)|_{L_2(\Omega)} \leq \frac{C\, h^{2(1-\varepsilon)}}{t^{\frac{3}{4}}}\, e^{-\omega_0 t}\, |y_0|_{L_2(\Omega)};$$

$$|y_h^*(t) - y^0(t)|_{L_2(\Omega)} \leq C\, h^{\frac{1}{2}-\varepsilon}\, e^{-\omega_0 t}\, |y_0|_{L_2(\Omega)},$$

where $Y_h^*(t)$ is the solution to the original heat equation (6.1) with feedback boundary conditions $u(t) = -\frac{\partial}{\partial\nu} P_h y_h(t)$ in (6.1c). ∎

Remark 10.2. The rates of convergence provided by Theorem 10.1 are optimal in the sense that they reconstruct the optimal regularity of the original solutions. The presence of factor $\frac{1}{t^{1-\varepsilon}}$ in Part III is consistent with 'rough data estimates' available for parabolic semigroups. ∎

Remark 10.3. To emphasize a contrast, we shall now consider the heat equation (6.1a–b) with Neuman control (6.12) and cost (6.22) where then $Y = L_2(\Omega)$ (rather than cost (6.13)) where then $Y = H^1(\Omega)$. The detailed approximation treatment is given in [Las.6]. Now since $\gamma < \frac{1}{2}$, one can take the usual Galerkin approximations for A_h and singly define $B_h = \Pi_h B$. This choices produce optimal rate $\mathcal{O}(h^{2(1-\epsilon)})$ of convergence with any spline approximation of order $s \leq 2$. ∎

10.2. Class (H.1): The structurally damped plate problem in Example 3.1

We return to Example 6.1, model (6.25) in Section 6, with $n = \dim \Omega \leq 3$.

Choice of V_h. We shall select the approximating space $V_h \subset H^2(\Omega) \cap H_0^1(\Omega)$ to be a space of splines (e.g., cubic splines), which comply with the usual approximation properties

$$|Q_h z - z|_{H^\ell(\Omega)} \leq C h^{s-\ell} |z|_{H^s(\Omega)}, \quad z \in H^s(\Omega) \cap H_0^1(\Omega),$$

$$0 \leq \ell \leq 2; \ \ell \leq s < r; \quad (10.31)$$

$$|z_h|_{H^\alpha(\Omega)} \leq C h^{-s} |z_h|_{H^{\alpha-s}(\Omega)}, \quad 0 \leq \alpha \leq 2, \quad (10.32)$$

where Q_h is the orthogonal projection of $L_2(\Omega)$ onto V_h and where r is the order of approximation.

Choice of A_h. We let

$$A_h = Q_h A Q_h : V_h \to V_h;$$

i.e.,

$$(A_h \phi_h, \Psi_h)_\Omega = (\Delta\phi_h, \Delta\Psi_h)_\Omega = (A^{\frac{1}{2}}\phi_h, A^{\frac{1}{2}}\Psi_h)_\Omega, \quad \phi_h, \Psi_h \in V_h, \quad (10.33)$$

$$|A_h^{\frac{1}{2}}\phi_h|_{L_2(\Omega)} = |A^{\frac{1}{2}}\phi_h|_{L_2(\Omega)}; \quad |A_h^{\frac{1}{2}}\phi_h| \sim |\phi_h|_{H^2(\Omega)}, \quad \phi_h \in V_h, \quad (10.34)$$

where (10.34) is a consequence of (10.33). From elliptic estimates

$$\begin{cases} |(A^{-1}-A_h^{-1}Q_h)z|_{H^2(\Omega)} \leq C\, h^2 |z|_{L_2(\Omega)}\,; \\ |(A^{-\frac{1}{2}}-A_h^{-\frac{1}{2}}Q_h)z|_{H^2(\Omega)} \leq C\, h^2 |z|_{H^2(\Omega)}\,, \quad z \in \mathcal{D}(A^{\frac{1}{2}})\,. \end{cases} \tag{10.35}$$

Choice of A_h and B_h. To begin with, we let $Y_h \equiv V_{h1} \times V_{h2}$, where V_{h1} consists of the elements of V_h equipped with norm $|v_h|_{V_h^1} = |A_h^{\frac{1}{2}} v_h|_{L_2(\Omega)}$ and V_{h2} consists of the elements of V_h equipped with the $L_2(\Omega)$-norm. We shall write $x_h = [x_{h1}, x_{h2}] \in Y_h$. Next, we define

$$A_h: Y_h \to Y_h: A_h = \begin{vmatrix} 0_1 & Q_h \\ -A_h & -A_h^{\frac{1}{2}} \end{vmatrix}; \tag{10.36}$$

$$B_h: L_2(\Gamma) \to Y_h: B_h u = \begin{vmatrix} 0 \\ \mathcal{B}_h u \end{vmatrix}, \quad (\mathcal{B}_h u, v_h)_\Omega = v_h(x^0)u. \tag{10.37}$$

Finally, we let $\Pi_h: Y \to Y_h$ be defined as

$$\Pi_h = \begin{vmatrix} Q_h & 0 \\ 0 & Q_h \end{vmatrix}.$$

Computation of adjoints A_h^* and B_h^*. To compute the adjoints of A_h and B_h, we use the inner products generated by the topology on V_{h1} and V_{h2}. We find, as in the continuous case,

$$A_h^* = \begin{vmatrix} 0 & -Q_h \\ A_h & -A_h^{\frac{1}{2}} \end{vmatrix}; \quad B_h^* x_h = x_{h2}(x^0)$$

as it follows from $(A_h x_h, y_h)_{Y_h} = (x_h, A_h^* y_h)_{Y_h}$ and $(B_h u, x_h)_{Y_h}$ and $(u, B_h^* x_h)_U$, respectively.

Approximating control problem. With the above notation, the approximating dynamics (9.22) is now

$$\begin{cases} (\ddot{y}_h, \phi_h) + (A_h y_h, \phi_h) + (A_h^{\frac{1}{2}} y_h, \phi_h) = \phi_{h2}(x^0)u; \\ (A_h y_h, \phi_h) = (\Delta y_n, \Delta \phi_h), \quad \text{all } \phi_h \in V_h; \\ (y_h(0), \phi_h) = (y_0, \phi_h); \quad (\dot{y}_h(0), \phi_h) = (y_1, \phi_h), \end{cases} \qquad (10.38)$$

where all inner products are in $L_2(\Omega)$.

The optimal feedback control for the finite-dimensional problem is given by $u_h^0(t) \equiv -[\overline{P}_{h2}\psi_h(t)][x^0]$, where

$$P_h y_h \equiv \begin{cases} P_{h1} y_{h1} + P_{h2} y_{h2} \equiv \overline{P}_{h1} y_h; \\ P_{h3} y_{h1} + P_{h4} y_{h2} \equiv \overline{P}_{h2} y_h, \end{cases} \qquad (10.39)$$

and P_h satisfies the following algebraic equation with $L_2(\Omega)$-inner products

$$-(A_h x_{h2}, \overline{P}_{h1} y_h) + (A_h x_{h1} - A_h^{\frac{1}{2}} x_{h2}, \overline{P}_{h2} y_h) - (A_h \overline{P}_{h1} x_h, y_{h2})$$

$$+ (\overline{P}_{h2} x_h, A_h y_{h1} - A_h^{\frac{1}{2}} y_{h2}) + (A_h x_{h1}, y_{h1}) + (x_{h2}, y_{h2})$$

$$= (\overline{P}_{h2} x_h)(x^0)(\overline{P}_{h2} y_h)(x^0). \qquad (10.40)$$

Verification of assumptions of Theorems 9.2. In order to apply Theorems 9.2 and 9.3, we need to verify the approximating assumptions (A.1) = (9.13) through (A.6) = (9.18), as well as assumptions (9.31), (9.32). Indeed, the last two are plainly satisfied: (9.31) since $R = I$, while (9.32) follows from (6.30) and the argument below it (in essence, $A^{-(\frac{1}{2}-\varepsilon)}\delta \in L_2(\Omega)$, while $A^{-\varepsilon}$ is compact on $L_2(\Omega)$.

Assumption (A.1) = (9.13). This follows by applying the arguments of [C-T.2] of the continuous case to the finite-dimensional operator given by (10.36).

Assumption (A.2) = (9.14). By (10.35), we have that (A.2) with $s = 2$ holds true:

$$|(A_h^{-1} - A^{-1})x_h|_Y = |-(A^{-\frac{1}{2}} - A_h^{-\frac{1}{2}})x_{h1} + (A^{-1} - A_h^{-1})x_{h2}|_{H^2(\Omega)}$$

$$\leq C h^2 [|x_{h1}|_{H^2(\Omega)} + |x_{h2}|_{L_2(\Omega)}] = C h^2 [|x_h|_Y].$$

The same result holds for the adjoint A^*, in view of its definition.

Assumption (A.3) = (9.15). By Sobolev embedding and the inverse approximation property (10.32), we have for any $\varepsilon > 0$,

$$|B^* x_h|_U = |x_{h2}(x^0)| \leq C|x_{h2}|_{H^{n/2+\varepsilon}(\Omega)} \leq C h^{-n/2-\varepsilon}|x_{h2}|_{L_2(\Omega)}$$

$$\leq C h^{-n/2-\varepsilon}|x_h|_Y ,$$

and (A.3) follows since $\gamma s = (\frac{n}{4}+\varepsilon)2 > \frac{n}{2}$.

Assumption (A.4) = (9.16). By (10.31) we compute

$$|B^*(\Pi_h x - x)|_U = |(Q_h x_2)(x^0) - x_2(x^0)|_{R^1} \leq C|Q_h x_2 - x_2|_{H^{n/2+\varepsilon}(\Omega)}$$

$$\leq C h^{2-n/2-\varepsilon}\|x_2\|_{H^2(\Omega)} \leq C h^{2-n/2-\varepsilon}\|x\|_{\mathcal{D}(A^*)} .$$

Since $2(1-\gamma) = 2(1-\frac{n}{4}-\varepsilon) < 2-\frac{n}{2}-\varepsilon$, and (A.4) is satisfied.

Assumption (A.5) = (9.17). It coincides with (A.4).

Assumption (A.6) = (9.18). We compute

$$\|B^*\Pi_h x\|_U = \|x_{h_2}(x^0)\|_{L_2(\Omega)} \leq C\|x_{h_2}\|_{H^{n/2+\varepsilon}(\Omega)} \leq C\|x_h\|_{\mathcal{D}(A^*\gamma)}$$

as in [C-T.4], $\mathcal{D}(A^*\gamma) \subset H^{4\gamma}(\Omega) \times H^{2\gamma}(\Omega)$ and $2\gamma = 2(\frac{n}{4}+\varepsilon) = \frac{n}{2}+2\varepsilon > \frac{n}{2}+\varepsilon$.

Thus, we have verified all the assumptions of Theorem 9.2 and 9.3. Thus, Theorem 9.2 applies to our problem and yields the following convergence results:

(i) $$\|P_h\Pi_h - P\|_{\mathcal{L}(H^2(\Omega) \times L_2(\Omega))} \leq C h^{\varepsilon_0}, \quad \varepsilon_0 < \frac{4-n}{2};$$

(ii) $$\left\|(P_h\Pi_h - P)|_{x=x^0}\right\|_{\mathcal{L}(H^2(\Omega) \times L_2(\Omega); R^1)} \to 0 \quad \text{as } h\downarrow 0;$$

or equivalently

$$\left\| B^* P_h \Pi_h - B^* P \right\|_{\mathcal{L}(H^2(\Omega) \times L_2(\Omega); R)} \to 0 \quad \text{as } h \downarrow 0,$$

where P_h is computed from (10.40).

(iii) $\displaystyle\sup_{t \geq 0} e^{\overline{\omega}_p t} \|u_h^0(t) - u^0(t)\|_{\mathcal{L}(H^2(\Omega) \times L_2(\Omega); R)} \leq C h^{\varepsilon_0};$

$\displaystyle\sup_{t \geq 0} t^\varepsilon e^{\overline{\omega}_p t} \|y_h^0(t) - y^0(t)\|_{\mathcal{L}(H^2(\Omega) \times L_2(\Omega))} \leq C h^{\varepsilon_0}, \quad \varepsilon_0 < \frac{4-n}{2}.$

Application of Theorem 9.3 to our problem yields the following result: Let $u_h^*(t)$ be a feedback law given by

$$u_h^*(t) = -[\overline{P}_{h2} y(t)][x^0]$$

which we insert into the *original* dynamics (6.25) to obtain

$$w_{tt} + \Delta^2 w - \rho \Delta w_t = \delta(x - x^0) u_h^*; \quad w|_\Gamma = \Delta w|_\Gamma = 0.$$

Then the corresponding feedback system is uniformly (in h) exponentially stable in the topology of $H^2(\Omega) \times L_2(\Omega)$ and uniformly approximates the original feedback dynamics. This means that the numerical algorithm provides a feedback control which yields uniform (in h) stability results for the original system.

We conclude this section by pointing out that the other examples of Section 6 dealing with structurally damped plate problems can be dealt with by a similar approximating scheme, see [L-T.19].

10.3. Class (H.2): The wave equation with Dirichlet boundary control

In this subsection, we verify the applicability of the approximating Theorem 9.9 to the wave equation with $L_2(\Sigma)$-Dirichlet boundary control, treated in Section 7.1, Part I, Eq. (7.1).

Choice of V_h. Let $V_h \subset H_0^1(\Omega)$ be an approximating subspace of $L_2(\Omega)$. Let Q_h be the L_2-orthogonal projection of $L_2(\Omega)$ onto V_h. We assume that V_h enjoys the following approximating properties:

$$
\begin{cases}
\text{(i)} & |Q_h z - z|_{H_0^\alpha(\Omega)} \to 0, \quad z \in H_0^\alpha(\Omega), \quad |\alpha| \leq 1; \\[2ex]
\text{(ii)} & |\frac{\partial}{\partial \nu} z_h|_{L_2(\Omega)} \leq C\, h^{-\frac{1}{2}-\epsilon} |z_h|_{H_0^1(\Omega)}; \\[2ex]
\text{(iii)} & |z_h|_{H_0^\alpha(\Omega)} \leq C\, h^{-s} |z_h|_{H_0^{\alpha-s}(\Omega)}, \quad 0 \leq \alpha \leq 1, \ \alpha-s \geq 0; \\[2ex]
\text{(iv)} & |\frac{\partial}{\partial \nu}(Q_h z - z)|_{L_2(\Omega)} \leq C\, h^{\frac{1}{2}} |z|_{H^2(\Omega)}.
\end{cases}
\tag{10.41}
$$

It is well known that the above approximation properties are satisfied for, say, spline approximations (of order r, r ≥ 1) defined on a uniform mesh. Also, modal (eigenfunction), or spectral (polynomial) approximations are typical examples of schemes which satisfy the requirements in (10.41).

<u>Choice of A_h.</u> We let

$$
A_h = Q_h A Q_h: V_h \to V_h; \quad \text{i.e.,} \quad (A_h \phi_h, \psi_h) = \int_\Omega \nabla \psi_h \cdot \nabla \psi_h \, d\Omega.
$$

From the definition of A_h, it follows immediately that

$$
|A_h^{\frac{1}{2}} z_h|_{L_2(\Omega)} = |A^{\frac{1}{2}} z_h|_{L_2(\Omega)},
$$

and $A_h^{\frac{1}{2}}$ is an isomorphism $H_0^1(\Omega) \to L_2(\Omega) \cap V_h$ (with a norm uniform in h). Since $A_h^{\frac{1}{2}}$ is self-adjoint on $L_2(\Omega)$, we have that $A_h^{\frac{1}{2}}$ is also an isomorphism $L_2(\Omega) \to H^{-1}(\Omega) \cap V_h$. Moreover, elliptic estimates give

$$
|(A_h^{-1} Q_h - A^{-1}) z|_{H_0^s(\Omega)} \leq C\, h^{2-s} |z|_{L_2(\Omega)}, \quad 0 \leq s \leq 1.
\tag{10.42}
$$

<u>Choice of A_h and B_h.</u> We introduce the space $Y_h = V_{h1} \times V_{h2}$, where V_{h1} (resp. V_{h2}) is the space V_h equipped with norm $L_2(\Omega)$ (resp. $A_h^{-\frac{1}{2}}$-norm):

$$
|v_h|_{V_{h1}} = |v_h|_{L_2(\Omega)}; \quad |v_h|_{V_{h2}} = |A_h^{-\frac{1}{2}} v_h|_{L_2(\Omega)}.
$$

By (10.42), Y_h is topologically equivalent to Y. The operators A_h and $B_h: U \to Y_h$ are defined by

$$A_h = \begin{vmatrix} 0 & I \\ A_h & 0 \end{vmatrix}; \; B_h u = \begin{vmatrix} 0 \\ \mathcal{B}_h u \end{vmatrix} = \begin{vmatrix} 0 \\ Q_h \mathcal{A} Du \end{vmatrix}, \tag{10.43}$$

well defined since

$$(\mathcal{B}_h u, \phi_h)_{L_2(\Omega)} = (\mathcal{A} Du, \phi_h)_{L_2(\Omega)} = \left[u, \frac{\partial \phi_h}{\partial \nu}\right]_{L_2(\Omega)}. \tag{10.44}$$

Finally, we let $\overline{\Pi}_h : Y_h \to Y_h$ be as before, below (10.37): a diagonal matrix with Q_h on the main diagonal.

Computation of adjoints A_h^* **and** B_h^*. These are computed with respect to the topologies of V_{h1}, V_{h2}. As in the continuous case, one readily obtains

$$A_h^* = -A_h \text{ and } B_h^* x_h = \frac{\partial}{\partial \nu} A_h^{-1} x_{h2}. \tag{10.45}$$

Approximating control problem. With the above notation, the discrete version of the state equation is equivalent to

$$\begin{cases} (\ddot{y}_h(t), \phi_h)_\Omega + \int_\Omega \nabla y_h \cdot \nabla \phi_h \, d\Omega = \left[u(t), \frac{\partial}{\partial \nu} \phi_h\right]_\Gamma; \\ (y_h(0), \phi_h)_\Omega = (Y_0, \phi_h)_\Omega, \; (\dot{y}_h(0), \phi_h)_\Omega = (Y_1, \phi_h)_\Omega, \end{cases} \tag{10.46}$$

for all $\phi_h \in V_h$, in the L_2-inner products of Ω and Γ, where

$$y_h(t) = \sum_{\ell=1}^h y_\ell \phi_\ell(x) = y_{h1}; \; y_{h2}(t) = \dot{y}_h(t).$$

The optimal feedback control for the finite-dimensional problem (9.22) is given by

$$u_h^0(t) = -\frac{\partial}{\partial \nu} A_h^{-1} \overline{P}_{h2}(y_h(t)) \tag{10.47a}$$

$$P_h \begin{bmatrix} x_{h1} \\ x_{h2} \end{bmatrix} = \begin{bmatrix} P_{h1} x_{h1} + P_{h2} x_{h2} \\ P_{h3} x_{h1} + P_{h4} x_{h2} \end{bmatrix} \equiv \begin{bmatrix} \overline{P}_{h1}(x_h) \\ \overline{P}_{h2}(x_h) \end{bmatrix}, \tag{10.47b}$$

and P_h satisfies the Riccati equation

$$-(\overline{P}_{h1} x_h, y_{h2})_\Omega + (\overline{P}_{h2} x_h, y_{h1})_\Omega - (x_{h2}, \overline{P}_{h1} y_h)_\Omega$$

$$+ (x_{h1}, \overline{P}_{h2}y_h)_\Omega + (x_{h1}, y_{h1})_\Omega + (A_h^{-1}x_{h2}, y_{h2})_\Omega$$

$$= \left[\frac{\partial}{\partial \nu} A_h^{-1}\overline{P}_{h2}, \frac{\partial}{\partial \nu} A_h^{-1}\overline{P}_{h2}y_h \right]_\Gamma ,$$

for all $x_h = \begin{bmatrix} x_{h1} \\ x_{h2} \end{bmatrix}$; $y_h = \begin{bmatrix} y_{h1} \\ y_{h2} \end{bmatrix} \subset V_{h1} \times V_{h2} = Y_h.$ (10.48)

Thus, if one can apply the theory presented in Theorem 9.9, then the feedback law (10.47) computed with the aid of Riccati operator P_h (from (10.48)), once inserted into the dynamics (10.46) yields a feedback semigroup which is uniformly exponentially stable, and it is convergent to the original, infinite-dimensional solution.

It remains to verify the assumptions of Theorem 9.9; i.e., conditions (B.1) = (9.50) through (B.4) = (9.53), (F.C.C.)$_h$ = (9.65), and (D.C.)$_h$ = (9.66)-(9.67), since the other assumptions (H.2) = (1.6), (F.C.C.) = (1.9), and (D.C.) = (5.17)-(5.10) have already been verified in Part I, Section 7.1.

Assumptions (9.1), (B.1), (B.2), (B.4), (D.C.)$_h$. By using the approximating properties (10.41), together with the elliptic estimates (10.42), one can show ([Las.3]) that hypotheses (B.1), (B.2), (B.4) are indeed satisfied for our dynamics of the wave equation with Dirichlet control (7.1). Indeed, assumption (D.C.)$_h$ = (9.66)-(9.67) is automatically satisfied, since R = I in our case.

Assumption (9.1).

$$|\Pi_h x - x|_Y \leq |Q_h x_1 - x_1|_{L_2(\Omega)} + |A^{-\frac{1}{2}}[Q_h x_2 - x_2]|_{L_2(\Omega)},$$

and since the inverse approximation holds, we have

$$|Q_h x - x|_{\mathcal{D}(A^\alpha)} \to 0, \quad x \in \mathcal{D}(A^\alpha), \quad |\alpha| \leq \frac{1}{2}.$$

which implies the desired conclusion.

Assumptions (B.1) = (9.50), (B.2) = (9.51). Let $z_h(t) \equiv e^{A_h t} x_h$; $x_h = \Pi_h x$; $x = (x_1, x_2)$. Then $z_h(t)$ satisfies:

$$\dot{z}_{h,1}(t) = -z_{h,2}(t)$$

$$(\ddot{z}_{h,1}(t), \phi_h)_\Omega + (\nabla z_{h,1}(t), \nabla \phi_h)_\Omega = 0,$$

or equivalently,

$$(\ddot{z}_{h,1}(t), \phi_h) + (A_h z_{h1}(t), \phi_h)_\Omega = 0.$$

Take $\phi_h \equiv A_h^{-1} \dot{z}_{h,1}(t)$. We have

$$\frac{1}{2} \frac{d}{dt} [A_h^{-\frac{1}{2}} \dot{z}_{n,1}(t)|^2_{L_2(\Omega)} + |z_{h,1}(t)|^2_{L_2(\Omega)}] = 0,$$

which gives

$$|\dot{z}_{h,2}(t)|^2_{H^{-1}(\Omega)} + |z_{h,1}(t)|^2_{L_2(\Omega)} = |\dot{z}_{h,1}(t)|_{H^{-1}(\Omega)} + |z_{h,1}(t)|^2_{L_2(\Omega)}$$

$$\leq |q_h x_2|_{H^{-1}(\Omega)} + |q_h x_1|_{L_2(\Omega)},$$

and proves (B.2) = (9.51). As for (B.1) = (9.50), we have

$$|A_h^{-1} \Pi_h x - A^{-1} x|_Y \leq |A_h^{-1} \Pi_h x - \Pi_h A^{-1} x|_Y + |(\Pi_h - I) A^{-1} x|_Y.$$

Convergence to zero of the second form on the right hand side of the above expression follows from the convergence properties of Π_h. It remains to estimate that

$$|A_h^{-1} \Pi_h x - \Pi_h A^{-1} x|_Y = |(A_h^{-1} Q_h - A^{-1}) x_2|_{L_2(\Omega)}$$

$$= |(A_h^{-1} Q_h - A^{-1}) Q_h x_2|_{L_2(\Omega)} + |A^{-1}(Q_h x_2 - x_2)|_{L_2(\Omega)}$$

$$\leq C h^2 |Q_h x_2|_{L_2(\Omega)} + |Q_h x_2 - x_2|_{H^{-1}(\Omega)}$$

$$\leq C h |Q_h x_2|_{H^1(\Omega)} + |Q_h x_2 - x_2|_{H^{-1}(\Omega)} \to 0,$$

where we have used (10.42), the inverse approximation property, and convergence properties of Q_h on $H^{-1}(\Omega)$.

Since $A_h^* = -A_h$, the proof of (B.1) for A_h^* is the same.

Assumption (B.4) = (9.53).

Part (i) = (9.54).

$$A^{-1}(B_h-B)u = \begin{bmatrix} 0 & A^{-1} \\ -I & 0 \end{bmatrix} \begin{bmatrix} 0 \\ (Q_h-I)ADu \end{bmatrix} = \begin{bmatrix} A^{-1}(Q_h-I)ADu \\ 0 \end{bmatrix}$$

$$|A^{-1}(B_h-B)u|_Y = |A^{-1}(Q_h-I)ADu|_{L_2(\Omega)}.$$

Denote $T_h \equiv A^{-1}(Q_h-I)AD$. Then the adjoint T_h^* of $T_h: L_2(\Gamma) \to L_2(\Omega)$ is given by

$$T_h^*z = \frac{\partial}{\partial\nu}(Q_h-I)A^{-1}z$$

and

$$|T_h^*z|_{L_2(\Gamma)} \le c\, h^{\frac{1}{2}}|A^{-1}z|_{H^2(\Omega)} \le c\, h^{\frac{1}{2}-\varepsilon}|z|_{L_2(\Omega)},$$

which proves that $T_h \to 0$, as desired.

Part (ii) = (9.55).

$$|(A_h^{-1}-A^{-1})B_h u|_Y = |(A_h^{-1}-A^{-1})\mathcal{B}_h u|_{L_2(\Omega)}.$$

Let $T_h \equiv (A_h^{-1}-A^{-1})\mathcal{B}_h: L_2(\Gamma) \to L_2(\Omega)$. Then

$$T_h^*z = \frac{\partial}{\partial\nu}Q_h(A_h^{-1}Q_h-A^{-1})z$$

and

$$|T_h^*z|_{L_2(\Gamma)} \le c\, h^{-\frac{1}{2}-\varepsilon}|Q_h(A_h^{-1}Q_h-A^{-1})z|_{H^1(\Omega)}$$

$$\le c\, h^{-\frac{1}{2}-\varepsilon}|(A_h^{-1}Q_h-A^{-1})z|_{H_0^1(\Omega)} \le c\, h^{-\frac{1}{2}-\varepsilon}h|z|_{L_2(\Omega)},$$

which, via duality, proves part (ii).

The proofs of part (iii) = (9.56) (resp. (iv) = (9.57)) follow (via transposition) from the proofs of part (i) (resp. (ii)). (Notice that in parts (i) and (ii), we have proved uniform convergence.)

Discussion on hypotheses (B.3) = (9.52) and (F.C.C.)$_h$ = (9.65). More delicate is the issue of validity of hypotheses (B.3) and (F.C.C.)$_h$ in the case of the wave equation problem with Dirichlet control (7.1) under study. Although both conditions are very natural--or they are

discrete counterpart of properties which hold true in the continuous
case--their validity in approximating problem (7.1) may well depend on
the specific selection of the numerical scheme adopted. This fact, or
pathology, should not--really--be surprising. Indeed, in the case,
say, of delay differential equations which are uniformly stabilizable
to begin with, it is known that the validity of the uniform Finite Cost
Condition (F.C.C.)$_h$ depends on the particular numerical scheme selected
(see [L-M], [M.1], [P]). Before we analyze further the validity of
conditions (B.3) and (F.C.C.)$_h$ in our present case of the wave equation
(7.1), it will be expedient to re-write them in an explicit, equivalent
form, as they apply to our case. It can be shown [Las.3] that

Lemma 10.1. Let $\psi_h(t) \in V_h$ be a semi-discrete solution of the
following ODE problem

$$(\ddot{\psi}_h(t), \phi_h)_\Omega + \int_\Omega \nabla\psi_h(t) \cdot \nabla\phi_h dr = 0, \; \forall \; \phi_h \in V_h. \tag{10.49}$$

Then:
(i)   Condition (B.3) = (9.52) is equivalent to the following
        inequality:

$$\left|\frac{\partial}{\partial\nu} \psi_h\right|_{L_2(\Sigma_T)} \leq c_T\left[|\dot{\psi}_h(0)|_{L_2(\Omega)} + \|\nabla\psi_h(0)\|_{L_2(\Omega)}\right]. \tag{10.50}$$

(ii)   Condition (F.C.C.)$_h$ = (9.65) is satisfied, provided

$$\left|\frac{\partial}{\partial\nu} \psi_h\right|_{L_2(\Sigma_T)} \geq c_T\left[|\dot{\psi}_h(0)|_{L_2(\Omega)} + \|\nabla\psi_h(0)\|_{L_2(\Omega)}\right]. \tag{10.51}$$

Remark 10.4. Notice that both condition (10.50), (10.51) are satisfied
for the **continuous** solutions of the homogeneous wave equation. In
fact, in this case, they are equivalent to, respectively, the
regularity assumption (H.2) = (1.5) and the exact controllability of
the wave equation with boundary control as in (7.1c), [L-T.1-3],
[Lio.1-2], [H.1], [T.3]. Moreover, the approximating subspace V_h
consists of polynomials defined on Ω, of degree $N(h) \to \infty$, as $h\downarrow0$, then
if ψ is such a polynomial, $\psi \in V_h$, then plainly $(x-x_0)\cdot\nabla\psi$ is also a
polynomial of degree N and hence $(x-x_0)\cdot\nabla\psi \in V_h$ as well. Thus the same

multiplier technique which yields the continuous counterpart of (10.50), (10.51) would give (10.50) and (10.51) as well. Thus we have

Lemma 10.3. Assume that the approximating space V_h consists of polynomials ψ of degree $N(h) \to \infty$ as $h\downarrow 0$ defined on Ω. Then inequalities (10.50) and (10.51) hold true. Consequently, the results of Theorem 9.9 apply in this case. ∎

To question whether these estimates (10.50), (10.51) are always satisfied for schemes which comply only with (10.41) is far from being obvious (see our earlier remark), and we are not in a position to give a full and general answer. It is clearly a technical issue which depends very much on the specific algorithm employed. Although very natural (in fact, necessary) for our problem, and although they have a continuous counterpart for the original system, their validity for a specific numerical scheme is, in general, an open problem of numerical analysis. Below, we shall provide an affirmative answer, but under an additional hypothesis (satisfied, for instance, by modal approxima-tions). The general case, say of spline approximations, is still an open question. (Here, we report some promising numerical computations [G-L-L], [D-G-K-W] which confirm numerically the validity of these conditions in the case of finite differences and mixed finite elements.)

Lemma 10.4. [Las.3] Assume that the following commutativity property holds true

$$Q_h A = A Q_h. \tag{10.52}$$

Then, any approximation scheme which complies with (10.41) and (10.42) satisfies the inequalities (10.50) and (10.51) as well. ∎

Remark 10.5. As noted above, examples of such an approximation are given, for instance, by modal approximations.

10.4. Class (H.2): First-order hyperbolic systems

In this section, following [Las.5], we shall verify the applicability of the approximating Theorem 9.8 on the Differential Riccati Equation, $T < \infty$ for the first-order hyperbolic problem of Section 7.4.

Choice of V_h. Let $\mathcal{V}_h \subset H^1(\Omega)$ be an approximating subspace of $L_2(\Omega)$ and let Q_h be the corresponding $L_2(\Omega)$-orthogonal projection. We assume that $\mathcal{V}_h$ enjoys the following approximation properties

$$|z|_{L_2(\Gamma)} \leq C \, h^{-\frac{1}{2}} |z_h|_{L_2(\Omega)} \, , \quad z_h \in \mathcal{V}_h; \tag{10.53}$$

$$|Q_h z - z|_{K^\ell(\Omega)} \leq C \, h^{r-\ell} |z|_{H^r(\Omega)} \, , \quad 0 \leq r \leq 2; \, 0 \leq \ell \leq 1. \tag{10.54}$$

Typical examples of the above approximating subspaces include spaces of spline approximations defined on a uniform grid. The approximation V_h of $Y = [L_2(\Omega)]^m$ and its corresponding projection Π_h are then

$$V_h = [\mathcal{V}_h]^m; \quad \Pi_h = [Q_h, \cdots, Q_h] \quad \text{(m-times).} \tag{10.54}$$

Choice of A_h. Let $A_h: V_h \to V_h$ be defined by

$$(A_h y_h, v_h)_Y = (F(x,\partial)y_h, v_h)_Y + (AD_1 My_h, v_h)_Y \, , \quad \forall \, y_h, v_h \in V_h \, ,$$

where F, A, D_1, M are defined in (7.63), (7.62), (7.64), (7.58c). By (7.63), (7.67), the above identity is equivalent to

$$(A_h y_h, v_h)_\Omega = (F(x,\partial)y_h, v_h)_\Omega + (My_h, A_N^- v_h^-)_U \, , \quad \forall \, y_h, v_h \in V_h \tag{10.55}$$

in the notation of Section 7.4, $U = [L_2(\Gamma)]^k$.

Remark 10.6. The subspaces V_h are not required to satisfy the boundary conditions; i.e., $V_h \not\subset \mathcal{D}(A)$. This is an important feature, since our framework requires a simultaneous approximation of both A and the adjoint A^*. Since the boundary conditions associated with $\mathcal{D}(A)$ and $\mathcal{D}(A^*)$ are not compatible, see (7.62), (7.69), the spaces V_h cannot be conformal. ∎

Choice of B. We take $B_h = \Pi_h B = \Pi_h(AD_1)$, i.e., by (7.67),

$$(B_h u, v_h)_\Omega = (u, D_1^* A^* v_h)_\Gamma = (u, A_N^- v_h^-)_\Gamma \, . \tag{10.56}$$

Computations of adjoints A_h^* and B_h^*. One readily obtains (see (5.14) in [Las.5])

$$(A_h^* y_h, v_h)_Y = (F^*(x, \partial) y_h, v_h)_Y - (A_N^+ y_h^+ + M^T A_N^- y_h^-, v_h^+)_U , \qquad (10.57)$$

where F^* is the formal adjoint of F, see (7.68). Also from (10.56) one obtains (see (7.67))

$$B_h^* v_h = B^* v_h = A_N^- v_h^- |_\Gamma . \qquad (10.58)$$

Approximating control problems. With the above notation, the discrete version of the state equation is equivalent to

$$\begin{cases} (y_h(t), \phi_h)_Y = (F(x, \partial) y_h(t), \phi_h)_Y + (M y_h - u, A_N^- \phi_h^-)_U , \\ (y_h(0), \phi_h)_Y = (y_0, \phi_h)_Y , \qquad \forall \, \phi_h \in V_h. \end{cases} \qquad (10.59)$$

The optimal feedback control for the finite dimensional problem is given by

$$u_h^0(t) = -A_N^- [P_h(t) y_h^0(t)]^- |_\Gamma , \qquad (10.60)$$

where $P_h(t)$ satisfies the DRE_h

$$(\dot{P}_h(t) y_h, v_h)_Y = -(R^* R y_h, v_h)_Y - (P_h(t) A_h y_h, v_h)_Y - (A_h^* P_h(t) y_h, v_h)_Y$$
$$+ (A_N^- [P_h(t) y_h]^-, A_N^- [P_h(t) y_h]^-)_U ,$$
$$P_h(T) = 0, \qquad \qquad \forall \, y_h, v_h \in V_h . \qquad (10.61)$$

Thus, in order to apply Theorem 9.8 on the DRE ($T < \infty$), we must verify the assumptions (B.1) = (9.50) through (B.4) = (9.53). As a matter of fact, we claim that these assumptions are satisfied, however, under the following additional hypothesis (which may in fact be relaxed):

$$\left.\begin{array}{l} \text{the matrices } A_j(\xi) \text{ are symmetric and block diagonal} \\ \\ A_j = \begin{vmatrix} A_j^- & 0 \\ 0 & A_j^+ \end{vmatrix} ; \quad j = 0, 1, 2, \cdots n \end{array}\right\} \qquad (10.62)$$

This assumption is automatically satisfied if dim Ω = 1. In fact, under an assumption weaker than (10.62) it is shown in [Las.5] that both A_h and A_h^* satisfy the following coercivity estimate for some $\alpha > 0$, which we write only for A_h:

$$-(A_h y_h, \hat{R} y_h)_Y \geq \alpha |y_h|_U^2 - c|y_h|_Y^2 . \qquad (10.63)$$

Here $\hat{R}$ is a certain invertible matrix (symmetrizer). These estimates for A_h and A_h^* are key elements in proving that assumptions (B.1) = (9.50) and (B.2) = (9.51) are then satisfied (see [Las.5; Theorem 5.1]). In effect (10.62) may be dispensed, as long as (10.63) can be obtained with an appropriate $\hat{R}$, and similarly for A_h^*. It is well known that in the continuous case such estimates always hold true [Kr.1] by using a pseudo-differential symmetrizer $\hat{R}$. On the other hand, in order to prove the stability estimate (B.2) = (9.51), one needs that $Ry_h \subset V_h$ which of course is not true if R is a pseudo-differential operator. Similarly, inequality (5.23) in [Las.5] asserts that

$$\left| e^{A_h^* t} y_h |_\Gamma \right|_{L_2(0,T;[L_2(\Gamma)]^k)} \leq c_T |y_h|_{[L_2(\Omega)]^m} . \qquad (10.64)$$

Inequality (10.64) combined with (10.58) yields the validity of assumption (B.3) = (9.52). Finally, the validity of (B.4) = (9.53ab) under the present assumption is established in [Las.5].

Conclusion. All required assumptions (B.1)-(B.4) are satisfied under the additional hypothesis (10.52). Thus, Theorem 9.8, Part I applies and yields in particular

$$\left| P_h(t)\Pi_h x - P(t)x \right|_{C([0,T];[L_2(\Omega)]^m)} \to 0, \quad x \in [L_2(\Omega)]^m, \qquad (10.65)$$

$$\left| y_h^0 - y^0 \right|_{C([0,T];[L_2(\Omega)]^m)} + \left| u_h^0 - u^0 \right|_{L_2(0,T;[L_2(\Gamma)]^k)} \to 0, \qquad (10.66)$$

where $P(t)$ is the viscosity solution of the DRE (3.21) given by (3.6).

In order to obtain a bonafide solution to the DRE and to claim convergence of the gain operators, we need to make additional assumptions on the observation operator R. Here we shall confine ourselves to note that the sufficient condition (9.63) of Part II of Theorem 9.8 is indeed satisfied in each of the cases 1 and 2 considered at the end of Section 7.4, where R is a finite rank bounded operator on Y and where R obeys hypothesis (7.75). We shall verify (9.63) with $R^*Rx = (x,c)c$: let $g \in C([0,T];Y)$

$$|\int_0^T B^*[e^{A_h^*t}\Pi_h - e^{A^*t}]R^*Rg(t)dt|_U = |\int_0^T B^*[e^{A_h^*t}\Pi_h - e^{A^*t}]c(g(t),c)_Y dt|_U^2$$

$$\leq \{\int_0^T |B^*[e^{A_h^*t}\Pi_h - e^{A^*t}]c|_U^2 dt\} \int_0^T |(g(t),c)_U|^2 dt. \qquad (10.67)$$

Convergence to zero of the first term in (10.67) as $h\downarrow 0$ is proved in [Las.5; Theorem 5.2]. Thus Theorem 9.8 Part II conclusion (9.64) on the gain operators holds true in our case with R finite rank. The same conclusion holds true for R satisfying (7.75), but proof is omitted.

In either case, $P_h(t)$ and $P(t)$ are bonafide solutions of DRE.

11.   Conclusions

The present two-part paper attempts to collect--within obvious space limitations--the most significant results concerning the existence (and uniqueness) of solutions to abstract operator Differential and Algebraic Riccati Equations as well as the numerical analysis of their numerical approximation for actual computations. We have considered a rather general dynamics (1.1) with unbounded operators B, subject to two different (but not mutually exclusive) regularity hypotheses, which were labeled (H.1) = (1.5) and (H.2) = (1.6) (first form) and $(H.2_R)$ = (1.8) (second form).

11.1.   Theoretical aspects

(H.1) class. This class comprises pairs $\{A,B\}$ of operators in (1.1), where A generates an s.c. analytic semigroup and B is "almost as unbounded as" A, in the technical sense of assumption (1.3).

Class (H.2). The regularity requirement (H.2) has been shown in recent years--by purely p.d.e. techniques--to hold true for a variety of hyperbolic problems as well as of plate problems with certain boundary conditions, and, moreover, for the Schrödinger equation with Dirichlet control, see the examples of Sections 4.1. However, there are still a few p.d.e. problems of physical importance where a gap still exists, at present, between the space where (H.2) is satisfied and the (smoother) space where exact controllability/uniform stabilization (hence the Finite Cost Condition (1.9)) have been ascertained. These cases are mentioned below.

In the case of _wave equations_ (or more generally, second order hyperbolic equations) with Neumann boundary control (as opposed to the Dirichlet control of section 4.1), the full theory of Theorem 5.2 applies in dim $\Omega = 1$ on the space $H^1(\Omega) \times L_2(\Omega)$. It is in relation to this space that assumption (H.2) holds true: This is the space of optimal regularity in this case of dim $\Omega = 1$, where exact controllability/uniform stabilization holds likewise true. However, for dim $\Omega \geq 2$, the situation is not so satisfactory, as mentioned in Remark 8.2. The solution to $L_2(\Sigma)$-controls lives pointwise in time in a space definitely larger than (with weaker topology than) the space $H^1(\Omega) \times L_2(\Omega)$ of finite "energy," see the sharp regularity results in [L-T.20-23], summarized in Remark 8.2, which moreover depend on the geometry of Ω. Thus, (H.2) is not true in relation to $H^1(\Omega) \times L_2(\Omega)$, if dim $\Omega \geq 2$. However, results of exact controllability/ uniform stabilization--needed to verify the Finite Cost Condition (1.9)--are presently available only on $H^1(\Omega) \times L_2(\Omega)$ with $L_2(\Sigma)$-controls, not in the spaces of sharp regularity where (H.2) instead is true.

In the case of _plates problems_, we have seen in section 7.2 that assumption (H.2) holds true for lower order boundary conditions, in relations to explicitly identified spaces of optimal regularity, where exact controllability/uniform stabilization results are also available to verify here the Finite Cost Condition. Thus, for these fourth-order problems (in space), the situation is as complete as for wave equations with Dirichlet control. However, for plates with higher order boundary conditions, the situation is, instead, rather the counterpart of wave equations with Neumann control. As in the latter case of waves, we have at present that assumption (H.2) is not true for plates with, say, shear forces and bending moments as controls in relation to the "energy" space $H^2(\Omega) \times L_2(\Omega)$, where instead exact controllability/uniform stabilization results are available [LL].

11.2. Numerical aspects

As to the numerical results discussed in this paper, we have shown that a fully consistent numerical theory can be developed for dynamical models which satisfy either assumption (H.1) = (1.5), or else (H.2) = (1.6).

Class (H.1). In the case of models which comply with assumption (H.1) ('analytic' class), the numerical theory is optimal, as it provides optimal convergence results (which are in line with the properties of the original continuous solutions) under <u>minimal</u> assumptions imposed on the approximating subspaces. These are just basic convergence properties which are satisfied by all spline approximations, modal or spectral approximations, etc.

Class (H.2). In the case of dynamics which comply with assumption (H.2), the numerical theory is less satisfactory. The basic theory of Theorem 9.2 is, yes, 'optimal,' in the sense that optimal convergence conclusions are obtained under only 'natural' approximating assumptions. However, the question remains as to whether or not a few of these 'natural' approximating assumptions are actually satisfied by various approximating subspaces (algorithms). This is, in fact, a delicate issue, and our Lemmas 10.3 and 10.4 provide just one answer. The issue is purely technical: whether a given approximating subspace will guarantee the desired (and 'natural') stability results as expressed by inequalities (10.50), (10.51). In full generality beyond these lemmas, this is presently an open question.

References

[B.1]   A. V. Balakrishnan, <u>Applied Functional Analysis</u>, Springer-Verlag, 2nd ed., 1981.

[B.2]   A. V. Balakrishnan, Boundary control of parabolic equations: L-Q-R theory, in <u>Non Linear Operators</u>, Proc. 5th Internat. Summer School, Akademie-Berlin, 1978.

[B-A]   I. Babuska and A. Aziz. <u>The Mathematical Foundations of the Finite Element Method with Applications to Partial Differential Equations</u>. Academic Press, New York, 1972.

[B-K]   T. H. Banks and K. Kunish, The Linear Regulator Problem for Parabolic Systems, <u>SIAM J. on Control</u>, Vol. 22, No. 5 (1984), 684-699.

[B-L-R.1] C. Bardos, G. Lebeau, and J. Rauch, Controle et stabilisation de l'equation des ondes, Appendix II in [Lio.3].

[B-L-R.2]   C. Bardos, G. Lebeau, and J. Rauch, Sharp sufficient conditions for the observation, control and stabilization of waves from the boundary

[B-T]   I. Bartolomeo and R. Triggiani, Uniform energy decay rates for Euler-Bernoulli equations with feedback operators in the Dirichlet/Neumann boundary conditions, SIAM _J. Mathem. Anal._, vol. 22 (1991), 46-71.

[B-S-T-W.1]   J. Bramble, A. Schatz, V. Thomee, and L. Wahlbin, Some convergence estimates for semidiscrete Galerkin type approximations for parabolic equations, SIAM _J. of Numerical Anal._, 14 (1977), 218-241.

[C-L]   S. Chang and I. Lasiecka, Riccati equations for non-symmetric and non-dissipative hyperbolic systems with L_2-boundary controls, _J. Math. Anal. and Appl._, Vol. 116 (1986), 378-414.

[C-P]   R. Curtain and A. Pritchard, Infinite dimensional linear systems theory, _LNCS_ 8, Springer-Verlag, 1978.

[C-R]   G. Chen and D. L. Russell, A mathematical model for linear elastic systems with structural damping, _Quarterly of Applied Mathematics_, January (1982), 433-454.

[C-T.1]   S. Chen and R. Triggiani, Proof of two conjectures of G. Chen and D. L. Russell on structural damping for elastic systems: The case $\alpha = \frac{1}{2}$, Proceedings of the Seminar in Approximation and Optimization held at University of Havana, Cuba, January 12-14, 1987, _Lecture Notes in Mathematics_, 1354, Springer-Verlag, 234-256.

[C-T.2]   S. Chen and R. Triggiani, Proof of extension of two conjectures on structural damping for elastic systems: The case $\frac{1}{2} \leq \alpha \leq 1$, _Pacific J. Mathematics_, Vol. 136, N1 (1989), 15-55.

[C-T.3]   S. Chen and R. Triggiani, Gevrey class semigroups arising from elastic systems with gentle dissipation: The case $0 < \alpha < \frac{1}{2}$, _Proc. Amer. Math. Soc._, vol. 110 (1990), 401-415.

[C-T.4]   S. Chen and R. Triggiani, Characterization of domains of fractional powers of certain operators arising in elastic systems, and applications, preprint 1989, _J. Diff. Eqns._, vol. 88 (1990), 279-293.

[DaP.1]   G. Da Prato, Quelques résultats d'existence, unicité, et régularité pour un probleme de la théorie du controle, _J. Math. Pures Appl._ 62 (1973), 353-375.

[DaP.2]   G. Da Prato, Lecture Notes, Scuola Normale Superiore, Pisa, 1990.

[D-I]   G. Da Prato and A. Ichikawa, Riccati equations with unbounded coefficients, _Ann. Matem. Pura e Appl._ 140 (1985), 209-221.

153

[DaP-L-T.1] G. Da Prato, I. Lasiecka, and R. Triggiani, A direct study of Riccati equations arising in boundary control problems for hyperbolic equations, *J. Diff. Eqns.*, Vol. 64, No. 1 (1986), 26-47.

[D-S] M. Delfour and Sorine, The linear-quadratic optimal control problem for parabolic systems with boundary control through the Dirichlet condition, 1982, Tolouse Conference, I.13-I.16.

[D-L-S.1] W. Doesch, I. Lasiecka, and W. Schappacher, Finite dimensional boundary feedback control problems for linear infinite dimensional systems, Israel J. of Math. 51 (1985), 177-207.

[D-G-K-W] T. Dupont, R. Glowinski, W. Kinton, M. Wheeler, Mixed finite element methods for time dependent problems: Application to control, Research Report UM/MD-54, Univ. of Houston, 1989.

[Fa.1] H. O. Fattorini, Boundary control systems, *SIAM J. Control*, vol. 6 (1968), 349-385.

[F.1] F. Flandoli, Riccati equation arising in a boundary control problem with distributed parameters, *SIAM J. Control and Optimiz.* 22 (1984), 76-86.

[F.2] F. Flandoli, Algebraic Riccati equation arising in boundary control problems, *SIAM J. Control and Optimiz.* 25 (1987), 612-636.

[F.3] F. Flandoli, A new approach to the LQR problem for hyperbolic dynamics with boundary control, Springer-Verlag, *LINCIS* 102 (1987), 89-111.

[F.4] F. Flandoli, Invertibility of Riccati operators and controllability of related systems, *Systems and Control Letters* 9 (1987), 65-72.

[F.5] F. Flandoli, On the direct solutions of Riccati equations arising in boundary control theory, Preprint di Matematica #66, March 1990, Scuola Normale Superiore, Pisa, Italy.

[F.6] F. Flandoli, A counterexample in the boundary control of parabolic systems, *Appl. Math. Letters*, to appear.

[F-L-T.1] F. Flandoli, I. Lasiecka, and R. Triggiani, Algebraic Riccati equations with non-smoothing observation arising in hyperbolic and Euler-Bernouli equations, *Ann. Matem. Pura a Appl.*, Vol. CLiii (1988), 307-382.

[Gib] J. S. Gibson, The Riccati integral equations for optimal control problems on Hilbert spaces, *SIAM J. Control and Optimiz.* 17 (1979), 537-565.

[G-A] J. S. Gibson and A. Adamian. Approximation Theory for the LQG Optimal Control of Flexible Structures, ICASE Report No. 88-48, 1988.

[G-L-L]   R. Glowinski, C. H. Li, J. L. Lions, A numerical approach
            to the exact boundary controllability of the wave equation,
            <u>Japan J. of Appl. Mathematics</u>, 7 (1990), 1-76.

[Gr]   P. Grisvard, Caracterization de qualques espaces
            d'interpolation, <u>Arch. Rational Mechanics and Analysis</u> 25
            (1967), 40-63.

[H.1]   L. F. Ho, Observabilite frontiere de l'equation des ondes,
            <u>CRAS</u>, Vol. 302, Paris (1986), 443-446.

[Hor.1]   M. A. Horn, Exact controllability of the Euler-Bernoulli
            plate via bending moments only on the space of optimal
            regularity, <u>J. Math. Anal. and Appl.</u>, to appear.

[Hor.2]   M. A. Horn, Uniform decay rates for the solutions of Euler-
            Bernoulli plate equation with boundary feedback acting via
            bending moments, University of Virginia, preprint 1991.

[I.1]   K. Ito. Strong convergence and convergence rate of
            approximating solutions for Algebraic Riccati equations in
            Hilbert spaces, <u>Lecture Notes Control Information Sciences</u>,
            102 (1987), Springer-Verlag.

[I-T]   K. Ito, M. T. Tran, Linear quadratic control problem for
            linear systems with unbounded input and output operators:
            Numerical approximations, Proceedings of the Vorau
            Conference, 1988.

[K.1]   H. O. Kreiss, Initial boundary value problems for
            hyperbolic systems, <u>Comm. Pure & Appl. Math.</u>, vol. 23
            (1970), 277-298.

[K-K.1]   M. Knoller and K. Kunish, Convergence rates for the
            feedback operators arising in the linear quadratic
            regulator problem.

[K-S]   F. Kappel and D. Salamon, An approximation theorem for the
            algebraic Riccati equation. Proceedings of the 27th IEEEC
            D.C. Conference in Austin, Texas, 1988.

[Lag.1]   J. Lagnese, Infinite horizon linear-quadratic regular
            problem for beams and plates, <u>Lecture Notes LNCIS</u>,
            Springer-Verlag, to appear.

[Lag.2]   J. Lagnese, Uniform boundary stabilization of homogeneous,
            isotropic plates, <u>Proc. of the 1986 Vorau Conference on
            Distributed Parameter Systems</u>, Springer-Verlag, 204-215.

[Las.1]   I. Lasiecka, Convergence estimates for semidiscrete
            approximations of nonselfadjoint parabolic equations, <u>SIAM
            J. Num. Anal.</u>, Vol. 21 (1984), 894-909.

[Las.2]   I. Lasiecka, Galerkin approximations of abstract parabolic
            boundary value problems with rough boundary data; L_p
            theory, <u>Math. Comp.</u>, Vol. 47, No. 175 (1986), 55-75.

[Las.3]   I. Lasiecka, Approximations of the solutions of infinite
            dimensional Algebraic Riccati Equations with unbounded
            input operators, <u>Numerical Funct. Anal. and Optimiz.</u>

[Las.4]   I. Lasiecka, Unified theory for abstract parabolic boundary
             problems; a semigroup approach, <u>Appl. Math. and Optimiz.</u>, 6
             (1980), 283-333.

[Las.5]   I. Lasiecka, Approximations of Riccati equations for
             abstract boundary control problems: Applications to
             hyperbolic systems, <u>Numerical Functional Anal. & Optimiz.</u>,
             Vol. 8; n. 3 and 4 (1985-86), 207-248.

[Las.6]   I. Lasiecka, Convergence rates for the approximations of
             solutions to algebraic Riccati equations with unbounded
             coefficients. The case of analytic semigroups, preprint
             1990.

[Las.7]   I. Lasiecka, Exponential decay rate for the Euler-Bernoulli
             equation with boundary description only in the moments.
             <u>J. Diff. Eqns.</u>, to appear.

[Leb.1]   J. Lebeau, Controle de l'equation de Schrödinger, preprint
             1989.

[Lio.1]   J. L. Lions, <u>Controle des Systems Distribues Singuliers</u>,
             Gauthier Villars, 1983.

[Lio.2]   J. L. Lions, Exact controllability, stabilization and
             perturbations, <u>SIAM Review</u> 30 (1988), 1-68.

[Lio.3]   J. L. Lions, <u>Controllabilite exacte, perturbations et
             stabilization de systemes distribues</u>, vols. 1 and 2, Masson
             1990.

[Lio-Mag.1] J. L. Lions and E. Magenes, <u>Nonhomogeneous boundary value
             problems</u>, Vol. I,II, Springer-Verlag, 1972.

[Lit.1]   W. Littman, Near optimal time boundary controllability for
             a class of hyperbolic equations, <u>Springer-Verlag Lecture
             Notes LNCIS</u> #97 (1987), 307-312.

[Lit.2]   W. Littman, talk at American Mathematical Society meeting,
             University of South Florida, Tampa, March 1991.

[L-L.1]   J. Lagnese and J. L. Lions, Modeling, analysis and control
             of thin plates, <u>Collection Recherches en Mathematiques
             Appliquees</u>, Vol. 6, Masson, Paris, 1988.

[L-L-T]   I. Lasiecka, J. L. Lions, and R. Triggiani, Non-homogeneous
             boundary value problems for second order hyperbolic
             operators, <u>J. Mathem. Pure et Appl.</u>, 65 (1986), 149-192.

[L-M.1]   I. Lasiecka and A. Manitius, Differentiability and
             convergence of approximating semigroups for retarded
             functional differential equations, <u>SIAM J. Numerical Anal.</u>
             25 (1988), 883-907.

[L-R]   D. Lukes and D. Russell, The quadratic criterion for
             distributed systems, <u>SIAM J. Control</u> 7 (1969), 101-121.

[L-T.1]   I. Lasiecka and R. Triggiani, A cosine operator approach to
             modeling $L_2(0,T;L_2(\Gamma))$-boundary input hyperbolic equations,
             <u>Appl. Math. and Optimiz.</u>, Vol. 7 (1981), 35-83.

[L-T.2]   I. Lasiecka and R. Triggiani, Regularity of hyperbolic
           equations under boundary terms, <u>Appl. Math. and Optimiz.</u>,
           Vol. 10 (1983), 275-286.

[L-T.3]   I. Lasiecka and R. Triggiani, A lifting theorem for the
           time regularity of solutions to abstract equations with
           unbounded operators and applications to hyperbolic
           equations, <u>Proc. Amer. Math. Soc.</u> 103, 4 (1988).

[L-T.4]   I. Lasiecka and R. Triggiani, Dirichlet boundary control
           problem for parabolic equations with quadratic cost:
           Analyticity and Riccati's feedback synthesis, <u>SIAM J.
           Control and Optimiz.</u> 21 (1983), 41-68.

[L-T.5]   I. Lasiecka and R. Triggiani, An L_2-Theory for the
           Quadratic Optimal Cost Problem of Hyperbolic Equations with
           Control in the Dirichlet B.C., Workshop on Control Theory
           for Distributed Parameter Systems and Applications,
           University of Graz, Austria (July 1982); <u>Lecture Notes
           LNICS</u>, Vol. 54, Springer-Verlag (1982), 138-153.

[L-T.6]   I. Lasiecka and R. Triggiani, Riccati equations for
           hyperbolic partial differential equations with
           $L_2(0,T;L_2(\Gamma))$-Dirichlet boundary terms, <u>SIAM J. Control and
           Optimiz.</u>, Vol. 24 (1986), 884-924.

[L-T.7]   I. Lasiecka and R. Triggiani, The regulator problem for
           parabolic equations with Dirichlet boundary control; Part
           I: Riccati's feedback synthesis and regularity of optimal
           solutions, <u>Appl. Math. and Optimiz.</u>, Vol. 16 (1987),
           147-168.

[L-T.8]   I. Lasiecka and R. Triggiani, The regulator problem for
           parabolic equations with Dirichlet boundary control; Part
           II: Galerkin approximation, <u>Appl. Math. and Optimiz.</u>, Vol.
           16 (1987), 198-216.

[L-T.9]   I. Lasiecka and R. Triggiani, Infinite horizon quadratic
           cost problems for boundary control problems, <u>Proceedings
           20th CDC Conference</u>, Los Angeles (December 1987),
           1005-1010.

[L-T.10]   I. Lasiecka and R. Triggiani, Differential Riccati
           Equations with unbounded coefficients: Applications to
           boundary control/boundary observation hyperbolic problems,
           <u>J. of Nonlinear Analysis</u>, to appear.

[L-T.11]   I. Lasiecka and R. Triggiani, Exact controllability of the
           Euler-Bernoulli equation with controls in Dirichlet and
           Neumann boundary conditions: A non-conservative case, <u>SIAM
           J. Control and Optimiz.</u>, 27 (1989), 330-373.

[L-T.12]   I. Lasiecka and R. Triggiani, Uniform exponential energy
           decay of wave equations in a bounded region with
           $L_2(0,\infty;L_2(\Gamma)$-feedback control in the Dirichlet boundary
           conditions, <u>J. Diff. Eqns.</u> 66 (1987), 340-390.

[L-T.13]   I. Lasiecka and R. Triggiani, Exact controllability of the wave equation with Neumann boundary control, **Applied Math. and Optimiz.** 19 (1989), 243-290.

[L-T.14]   I. Lasiecka and R. Triggiani, Regularity theory for a class of nonhomogeneous Euler-Bernoulli equations: A cosine operator approach, Bollett. **Unione Mathem. Italiana UMI** (7), 3-B(1989), 199-228.

[L-T.15]   I. Lasiecka and R. Triggiani, Exact controllability of the Euler-Bernoulli equation with boundary controls for displacement and moment, **J. Math. Anal. & Appl.**, 146 (1990), 1-33.

[L-T.16]   I. Lasiecka and R. Triggiani, Regularity, exact controllability and uniform stabilization of Kirchoff plates via only the bending moment, 1989, **J. Diff. Eqns.**, to appear.

[L-T.17]   I. Lasiecka and R. Triggiani, Stabilization and structural assignment of Dirichlet boundary feedback parabolic equations, **SIAM J. Control and Optimiz.**, vol. 21 (1983), 766-803.

[L-T.18]   I. Lasiecka and R. Triggiani, Stabilization of Neumann boundary feedback parabolic equations: the case of trace in the feedback loop, **Appl. Mathe. and Optimiz.**, vol. 10 (1983), 307-350.

[L-T.19]   I. Lasiecka and R. Triggiani, Numerical approximation for abstract systems modelled by analytic semigroups, and applications, Mathematics of Computation, to appear.

[L-T.20]   I. Lasiecka, and R. Triggiani, Sharp regularity results for mixed second order hyperbolic equations of Neumann type: The L_2-boundary case, **Ann. di Matem. Pura e Appl.**, to appear.

[L-T.21]   I. Lasiecka and R. Triggiani, Trace regularity of the solutions of the wave equation with homogeneous Neumann boundary conditions and compactly supported data, **J. Math. Anal. and Appl.**, vol. 141 (1989), 49-71.

[L-T.22]   I. Lasiecka and R. Triggiani, Riccati Differential Equations with unbounded coefficients and non-smooth terminal condition--The case of analytic semigroups, **SIAM J. Math. Anal.**, to appear.

[L-T.23]   I. Lasiecka and R. Triggiani, Regularity theory of hyperbolic equations with non-homogeneous Neumann boundary conditions. Part II: General boundary data, **J. Diff. Eqns.**, to appear.

[L-T.24]   I. Lasiecka and R. Triggiani, Recent advances in regularity of second order hyperbolic mixed problems, and applications, to appear in book series, **Dynamics Reported**, Wiley-Teubner.

[L-T.25]   I. Lasiecka and R. Triggiani, Uniform stabilization of the
            wave equation with Dirichlet feedback control without
            geometrical conditions, Appl. Math. and Optimiz., to
            appear.

[L-T.26]   I. Lasiecka and R. Triggiani, Optimal regularity, exact
            controllability and uniform stabilization of Schrödinger
            equations with Dirichlet control, Diff. and Integral Eqns.,
            to appear.

[L-T.27]   I. Lasiecka and R. Triggiani, Exact controllability and
            uniform stabilization of Euler-Bernoulli equations with
            boundary control only on $\Delta w|_\Sigma$, Bollettino Unione Matem.
            Italiano, to appear.

[L-T.28]   I. Lasiecka and R. Triggiani, Further results on exact
            controllability of Euler-Bernoulli equations with controls
            in the Dirichlet and Neumann boundary conditions,
            Springer-Verlag Lecture Notes in Control and Information
            Sciences, J. P. Zolesio, Editor, to appear; Proceedings
            International Workshop held in Montpellier, France, January
            1989.

[M.1]   Myatake, Mixed problems for hyperbolic equations of second
            order, J. Math. Kyoto Univ., vol. 130-3 (1973), 435-487.

[M-T-P-R]   A. Manitius, G. Tran, G. Payre, and R. Roy, Computation of
            eigenvalues associated with functional differential
            equations, SIAM J. Sci. Statist. Comp., to appear.

[M-T]   A. Manitius and R. Triggiani, Function space control-
            lability of linear retarded systems: A derivation from
            abstract operator conditions, SIAM J. Control, Vol. 16
            (1978), 599-645.

[N.1]   J. Nitsche, Uber ein Variationsprinzip zur Lösung von
            Dirichlet Problemen bei Verwendung von Teilräumen, die
            keinen Randbedingungen unterworfen sind, Abh. Math. Sem.
            Univ. Hamburg 36 (1971), 9-15.

[O-T]   N. Ourada and R. Triggiani, Uniform stabilization of the
            Euler-Bernoulli equation with feedback only on the Neumann
            B.C., preprint 1989, Diff. and Integral Eqns., vol. 4
            (1991), 277-292.

[P]   G. Propst, Numerische Untersuchung der Spectra Endlich-
            mensionaler Approximation für Differenzen-Differential-
            gleichungen, Technical Report No. 40, 1984, Univ. of Graz.

[P-P]   J. Pollock and A. Pritchard, The infinite time quadratic
            cost problem for distributed systems with unbounded control
            action, J. Inst. Math. Appl. 25 (1980), 287-309.

[P-S]   A. Pritchard and D. Salomon, The linear quadratic control
            problem for infinite dimensional systems with unbounded
            input and output operators, SIAM J. Control and Optimiz.,
            Vol. 25 (1987), 121-144.

[Rau.1]   J. Rauch, L_2 is a continuable initial condition for Kreiss'
           mixed problems, <u>Comm. Pure & Appl. Math.</u>, vol. 25 (1972),
           265-285.

[Ru.1]   D. Russell, Quadratic performance criteria in boundary
           control of linear symmetric hyperbolic systems, <u>SIAM J.
           Control</u> 11 (1973), 475-509.

[Ru.2]   D. Russell, Controllability and stabilizability theory for
           linear partial differential equations: recent progress and
           open questions, <u>SIAM Review</u> 20 (1978), 639-740.

[Sal]   D. Salamon, Infinite dimensional linear systems with
           unbounded control and observation: A functional analytic
           approach, <u>Trans. Am. Math. Soc.</u>, 1987, 383-431.

[Sh]   R. E. Showalter, Hilbert space methods for partial
           differential equations, Pitman, 1979.

[Sor]   M. Sorine, Une resultat d'existence et unicité pour
           l'equation de Riccati stationnaire, report <u>INRIA</u>, No. 55,
           1981.

[T.1]   R. Triggiani, On the stabilizability problem in Banach
           space, <u>J. Math. Anal. and Appl.</u>, Vol. 52 (1975), 383-403.
           Addendum J.M.A.A. 56 (1976), 492-3.

[T.2]   R. Triggiani, A cosine operator approach to modeling
           $L_2(0,T;L_2(\Gamma))$-boundary input problems for hyperbolic
           systems, Proceedings of 8th IFIP Conference, University of
           Wurzburg, W. Germany, July 1977, <u>Lecture Notes CIS</u>,
           Springer-Verlag #6 (1978), 380-390.

[T.3]   R. Triggiani, Exact boundary controllability on
           $L^2(\Omega) \times H^{-1}(\Omega)$ for the wave equation with Dirichlet control
           acting on a portion of the boundary, and related problems,
           <u>Appl. Math. and Optimiz.</u> 18 (1988), 241-277. Preliminary
           version in Springer-Verlag Lecture Notes in Control and
           Information Sciences, vol. 102 (1987), 291-332.

[T.4]   R. Triggiani, Regularity of structurally damped systems
           with point/boundary control, preprint 1989, <u>J. Math. Anal.
           and Appl.</u>, to appear.

[T.5]   R. Triggiani, Boundary feedback stabilizability of
           parabolic equations, <u>Appl. Math. and Optimiz.</u>, vol. 6
           (1980), 201-220.

[T.6]   R. Triggiani, On Nambu's boundary stabilizability problem
           of diffusion processes, <u>J. Diff. Eqns.</u>, vol. 33 (1979),
           189-200.

[T.7]   R. Triggiani, Uniform exponential energy decay of Euler-
           Bernoulli equations by suitable boundary feedback
           operators, Workshop held at Vorau, Austria, July 1988,
           International Series in Mathematics, vol. 91, Birkhauser
           (1989), 391-401.

[T.8]   R. Triggiani, Lack of exact controllability for wave and
          plate equations with finitely many boundary controls, _Diff._
          and Integral Eqns., to appear.

[T.9]   R. Triggiani, Exact Controllability for Wave and Euler-
          Bernoulli Equations in the Presence of Damping, 30 Years of
          Modern Optimal Control, _Lecture Notes in Pure and Applied_
          Mathematics, Vol. 119 Marcel Dekker (1989), 377-387.

[T.10]   R. Triggiani, Regularity with point control. Part I: Wave
          and Euler-Bernoulli equations, _Springer-Verlag Lecture_
          Notes, to appear.

[T.11]   R. Triggiani, Regularity with point control. Part II:
          Kirchhoff equations, February 1991.

[T.12]   R. Triggiani, Regularity with point control. Part III:
          Schrödinger equations, February 1991.

[T.13]   R. Triggiani, Wave equation on a bounded domain with
          boundary dissipation: An operator approach, _J. Math. Anal._
          & Appl., vol. 137 (1989), 438-461. Preliminary version in
          Lecture Notes in Pure and Applied Mathematics, vol. 108
          (1988), 283-310, Sung J. Lee, Editor.

[Ta.1]   D. Tataru, work in progress, University of Virginia, 1991.

[W]   D. Washburn, A bound on the boundary input map for
          parabolic equations with application to time optimal
          control, _SIAM J. Contr._ 17 (1979), 652-671.

[Z]   J. Zabczyk, Remarks on the algebraic Riccati equation in
          Hilbert space, _Appl. Math. and Optimiz._ 2 (1976), 251-258.

Lecture Notes in Control and Information Sciences

Edited by M. Thoma and A. Wyner

Lecture Notes in Control and Information Sciences

Edited by M. Thoma and A. Wyner

Lecture Notes in Control and Information Sciences

Edited by M. Thoma and A. Wyner